CLIMATE CHANGE

CLIMATE CHANGE

THE CONSEQUENCES OF THE CHANGING CLIMATE MAY STILL TAKE US BY SURPRISE!

MARGARET BANNAN

Copyright © 2021 by Margaret Bannan.

Library of Congress Control Number:		2021919932
ISBN:	Hardcover	978-1-9845-0847-8
	Softcover	978-1-9845-0846-1
	eBook	978-1-9845-0845-4

All rights reserved. No part of this book may be reproduced or transmitted in any form or by any means, electronic or mechanical, including photocopying, recording, or by any information storage and retrieval system, without permission in writing from the copyright owner.

Any people depicted in stock imagery provided by Getty Images are models, and such images are being used for illustrative purposes only. Certain stock imagery © Getty Images.

Print information available on the last page.

Rev. date: 10/12/2021

To order additional copies of this book, contact:
Xlibris
AU TFN: 1 800 844 927 (Toll Free inside Australia)
AU Local: (02) 8310 8187 (+61 2 8310 8187 from outside Australia)
www.Xlibris.com.au
Orders@Xlibris.com.au

833945

CONTENTS

Introduction xi

Chapter 1 Where Are We At in Our Understanding of
               Climate Change? 1

Chapter 2 Three Flashpoints for Alarm 7
- *Global Warming:* 8
- *Pollutants in the Atmosphere:* 10
- *Loss of Ecosystems and Biodiversity:* 11

Chapter 3 Demystifying Our Environmental History 13
- *Mystery of Life* 13
- *Hunters and Gatherers* 16
- *Agricultural Revolution* 18
- *Industrial Revolution* 19
- *Technological Revolution* 21
- *Environmental Crisis* 25

Chapter 4 What, in a Nutshell, Is Happening to the Biomes? ... 28
- *Tundra Biome* 29
- *Grasslands Biome* 33
- *Desert Biome* 38
- *Rainforest Biome* 41
- *Taiga Biome* 47

Chapter 5 What, in a Nutshell, Is Happening to Ecosystems? 53
- *Aquatic Ecology* 55
- *Ocean Ecology* 58
- *Air Ecology* 65
- *Soil Ecology* 69

- Antarctic Ecology 73
- Arctic Ecology 77
- Water Ecology 81
- Coral Reef Ecology 86
- Wetlands Ecology 91
- River Ecology 94
- Glacier Ecology 98
- Mangrove Ecology 102
- Artesian Ecology 106

Chapter 6   What, in a Nutshell, Is Happening to Biodiversity? 110
- Plant Biodiversity 111
- Insect Biodiversity 115
- Bird Biodiversity 120
- Animal Biodiversity 123
- Reptile Biodiversity 127
- Ocean Fish Biodiversity 130

Chapter 7   The Laborious Turning of the Anthropocene Era 136
- Anthropocene Era 136
- Ecological Period 140

Chapter 8   What Are We Thinking? 144
- Business as Usual 144
- Psychology of Denial 150
- Tipping Points 157
- Societal Chaos 162

Chapter 9   Philosophy of Our Ecology 170
- Everything Is Alive 170
- Ecopsychology 171
- Cosmology 173
- Earth Ecology 175
- Human Ecology 176
- Ecological Self 178
- Deep Ecology 180

Chapter 10 Assessing Change in Our Relationship with
            Nature..184
   - Ecological Footprint 185
   - Environmental Sustainability 187
   - Precautionary Principle 192

Chapter 11 Issues Worth a Good Look......................... 195
   - Cradle to Cradle 195
   - Carbon Sinks..197
   - Advent of Plastic 200
   - Population Explosion 204
   - Environmental Refugees 209
   - Plight of Bees... 212
   - Natural Disasters 214
   - Alternative Energy.. 217
   - Alternative Farming 220

Chapter 12 Spiritual Response 225
   - Science versus Religion..................................... 225
   - Science Invites All Religions to the Dance 226
   - Earth Spirituality 231
   - Spiritual Ecological Consciousness 233
   - Ecological Conversion...................................... 235

Chapter 13 We Are All in This Together 237
   - Join the Dance 237
   - A Way Forward 240
   - So what can I do?... 243

Chapter 14 Recapping the Endgame 247
   - Ambassadors for a Hospitable Earth........................ 247

Conclusion 253

This book is dedicated to climate scientists, our intellectual elite who hold the crystal ball that tells our future on Earth but who are not being heard. It is also dedicated to the youth of the world who are protesting for climate action. May the youth of the world be courageous and strong to persist indefinitely for urgent climate action.

INTRODUCTION

As I sit on my deck, I cannot see climate change. I cannot feel global warming; therefore, it is 'not happening'. This book is a call to arms for all who are sleepwalking in the dark to the possible extinction of the cleverest species on Earth, us *Homo sapiens*, wise humans. It is important to note the reason I am writing this book; it is because climate scientists are losing hope and faith in us to respond to the greatest challenge of our human existence. People in Australia are, per capita, amongst the biggest environmental consumers and polluters on Earth, and it will be to our eternal shame if we do not fight for the health of our home planet for our children and grandchildren. It is not rocket science to work out how we can turn global warming around; climate scientists have given us the roadmap because they have been there in their research and modelling. We have the knowledge, we have the intelligence, and we have the technology, but we lack the 'will' to make the changes for a safe future; this is what is most frustrating for climate scientists in Australia, and it is typical of climate scientists in other countries as well. A downside and upside of addressing global warming and environmental degradation is that the youth and children are leading the charge. Young people have taken up the challenge; they can't wait for us grown-ups to get it together in time to save us from a hostile Earth. All power, therefore, to the youth who are speaking out, striking, protesting, and rallying for us adults for a safer, more inhabitable Earth!

Throughout this book, I will refer to climate scientists as there are many scientists not involved in climate science who are perpetuating 'fake news', and they are quoted to allay people's fears so that governments can continue on their merry way with burning fossil fuels while totally ignoring the science of climate change as well as the loss of ecosystems and biodiversity with impunity. Climate scientists have been telling us for decades that if we continue the course we are on, then all life on Earth as we know it is in mortal peril. We have had half a century of indecision, procrastination, and denial that things could get so bad. Climate scientists are reluctant to tell us how bad things are because they don't want to scare us, and they get abused for being too dramatic, alarmist, and negative about our future.

We do not, in general, listen to the prophets in our midst. Perhaps we would rather listen to the shock jocks perpetrating and perpetuating a 'good news' story that human-induced climate change is just a conspiracy and that, like COVID-19, it will just go away. However, climate scientists hold the crystal ball that tells our future, and they are giving up on us. Some are planning not to have children because they know what the world is going to be like for their children, who will have to endure the consequences of unabated global heating or move their families to where they think they might have the best chance to survive. However, that will not be the answer because we know how people react when threatened, so no place on Earth will be safe. There are numerous hopeful environmental movements, but most of their efforts fall on deaf ears. It is like those in power are living in a dream world that all will be well if they ignore the problem. The islander people of the Pacific do not live that dream. They are already living the reality that their nation states are doomed by rising oceans that are heating up because of global warming. However, if you are reading this book, then you are predisposed to wanting to understand what is going on with our home planet.

We are normal people just doing our thing and going about our business as usual. We are people who live in countries that have enjoyed a most amazing lifestyle often removed from where the devastation of Earth is visible, but we are not immune from the effects of our behaviour. We are not climate scientists, but ignorance of what is happening to our home planet is not outside our understanding. Almost every day the news cycle presents us with an environmental news flash, and program after program on television, film after film, thousands of climate scientists' research documents, and hundreds of books on the changing climate, environmental degradation, and species extinctions are telling us what is going on with the health of Earth.

In my life, I am well acquainted with people who say they do not believe in human-induced climate change as though it is a matter of faith. Blind faith that it is all cyclical is not an option. The climate emergency is not something that you can choose to believe or not. It is a matter of mathematics, physics, chemistry, and now photography. We may not get all the science relating to the changing climate, but the camera does not lie. Documentary after documentary has presented real-life footage, images, photos, and films which clearly show what is happening to our home planet as it is transpiring. The triple FFF comes to mind – floods, fire, and famine – because of drought proving to be catastrophic for some countries and millions of people, but these tell-tale signs are just the beginning of our experience and education about the health of Earth. Environmental literacy is the most important literacy of all; we need to be able to read the signs of the Earth. So how Earth literate are we? We are earthlings; we live on this planet. What do we really know about how it functions as a planet? What are its planetary needs to be healthy? We know that Planet Earth is alive, it lives, and it has needs to stay healthy to support animate life.

The problem for climate scientists is that they struggle to communicate their research and predictions of possible outcomes regarding the changing climate to the general public. In general, we do not read scientific journals that inform us about their research. Climate scientists have tried to educate the public as best they can, even inviting religious leaders as they are the people who have access to a captive audience and can inform their attendees. All religions have within their belief and value systems the need to care for the Earth. However, religious leaders and preachers have let them down, so they have no place to turn to get their vital messages out. We are playing roulette with the future of biodiversity of life on Earth, and that includes us. If only we could just get our heads around the fact that we are just 'big bugs' on the Earth, members of the animal kingdom and integral members of the biosphere who are equally at risk from global heating of the planet!

As each scientist reveals something from their research, they may get a thirty-second window on the television, for example, but there is no continuity of the narrative that conditions for life on Earth are changing, and we need to be extremely attentive to those changes. We have hundreds of beautiful policy documents on climate change action and environmental degradation restoration, but there is not the degree of follow-up with real action that is so desperately needed. Money given in environmental grants seems to just disappear into the ether, and not much has been accomplished that has, in any way, averted our environmental emergency. The lack of political action in Australia and around the world has set us back ten years in mitigating the worst consequences of global warming.

To add insult to injury, many climate scientists have been silenced, gagged in the interests of logging, mining, and even water security for our people. The idea that we are on target to reach zero emissions in the second half of the century is a real worry as this will be too late. It is quite simply a matter of mathematics. If we pollute the air, heat the ocean, and destroy biodiversity at the same rate or greater over the next thirty years as we have over the last thirty years, then there will be no second half of the century to get it right. We will have given a climate runaway train a bright green light, and Mother Nature will be its captain. What does make for hard hearing for me (and I am sure for climate scientists as well) is that all through the COVID-19 pandemic, the constant refrain from governments was that they were listening to medical scientists and acting on their expert advice. Medical experts and governments came together to defend their people from a deadly virus. It was full steam ahead to flatten the curve of COVID-19. Why do we not listen to climate scientists and act on their expert advice on how to flatten the greenhouse curve? Increased global heating will be far worse than a virus pandemic, if that is possible.

CHAPTER ONE

Where Are We At in Our Understanding of Climate Change?

We hear a great deal about global warming and climate change, but understanding the accumulation of complexities associated with what they mean for us as individuals and the whole world community is a huge undertaking. However, what is needed is a critical mass of informed people to join together to bring pressure to bear on governments so that they will legislate for a carbon-free environment now if not sooner. All the inane dialogue about having until 2050 to get to zero emissions is insensitive to climate science. We know what thirty more years of pumping carbon dioxide, methane, and other greenhouse gases into the atmosphere will have on ecosystems and, more importantly, biodiversity because photographs show us the damage done by one degree of global heating.

Humans have never lived in a world like the one we are creating for ourselves, and common scientific understanding is that we may have already set ourselves up for an increase of warming of three degrees Celsius, which will be cataclysmic. Inaction now on global warming is going to have long-term devastating effects on all life, so it is no wonder climate scientists around the world are doing their best to inform us about the changing climate. Added to their research knowledge is the problem of not wanting to create eco-anxiety, which can seriously upset the mental health of people. Eco-anxiety for many people is a real possibility as people are confronted with not only an environmental emergency but also the potential for environmental catastrophe, which will impact on everyone. Climate scientists know that many of the human species in developed countries have had it so good that they will not be prepared to cope with the loss of lifestyle they are accustomed to; nor do they want to think about the impact of global warming on those they love. We have numerous examples of how people are affected, from experiencing never-ending drought to unprecedented bushfires that have scorched the Earth, their homes, and their livelihoods as well as extraordinary, devastating floods. To get a real picture of what climate scientists are experiencing, don't just listen to their words, their intelligent but nervous words; listen to their hearts. Don't just listen to their hearts; watch their body language as they speak the words we do not want to hear. Are we listening? Can we hear what they are saying? Can we find it in our hearts to be in empathy with them?

There are many environmental scientists who are weeping at the loss of their favourite nature interests, such as a local glacier guide who is watching glaciers disappearing before his eyes or a climate scientist who has spent his life researching and recording the loss of permafrost in Siberia. As the permafrost melts, he weeps not only for the loss of permafrost but also for the future of the world. Then there are marine scientists who have spent years documenting the submersion of islands through rising ocean levels, especially islands in the Pacific Ocean, or those recording the loss of coral reefs through bleaching. In listening to them, I have felt their sadness, loss, despair, hopelessness, anguish, fear, and pain. However, we cannot and must not let our own eco-anxiety be the reason for inaction. On the contrary, we have a very important job to do in turning around the mess that we have got ourselves into through ignorance, really. This is truly the great work that we as members of the human species are called to, and we must respond; our lives and everyone we love depend on it. It never crossed our minds that the human species could influence the ocean or the atmosphere, for example, but now we have to join with Mother Nature as she guides us to renew the Earth. COVID-19 has shown us how willing Mother Nature is to restore the balance and harmony of Planet Earth. While human 'big bugs' have been in lockdown, nature came out to play.

As members of the animal kingdom, we are very territorial – that is, we seem to be only interested in what is going on in our backyard and where our loved ones live. However, we are bushies, suburbians, townies, stateies, and countryites. What is required is that we understand first and foremost that we are earthlings. We are people who live within the total community of life on Planet Earth, which includes the rivers and seahorses. Whatever happens to the ice in the Arctic Circle will eventually affect us. Global warming is melting the ice. We are two-thirds water; we will not respond well if we get to three degrees Celsius of global heating. Heatwaves are responsible for more deaths than wildfires or cyclonic storms causing floods. The old, young, and vulnerable are most at risk because of the need to continually hydrate and stay cool. Whatever happens to the natural world will happen to us because we are not separate from the rest of the natural world. COVID-19 is a good example of our closeness to nature as a virus in animals can infect humans. We belong to the animal kingdom – not such a major jump for a zoonotic virus.

We are integrally connected and dependent on the conditions that allowed life to come into existence. Everything on Earth is interlinked; nothing and no one is separate from planetary life. The air we breathe here in downtown Melbourne, Australia, is the same air that prehistoric creatures breathed, and people who live on the icecaps will breathe after crocodiles and emus have exhaled it. Air is constantly recycled through nature's awesome recycling ecosystems; even the air over the Chernobyl and Fukushima nuclear disasters is part of the air we breathe, wherever we are on the planet. We are human earthlings along with the totality of creatures that depend on and share the same gifts of Earth. It is so important that we understand that we don't exist just within our own physical space. We live as integral creatures within Earth's biosphere. Whatever happens to the other-than-human natural world will happen to us because we are all interconnected through air and water. Our lives are completely intertwined with the rest of creation, which includes trees and ants.

There are some folks who are all for dismissing the now clear scientific evidence that something serious is going down with our home, Earth. Most ecosystems that support our lives are in imminent peril because of our activity, but that is, for some, just a myth. Some thoughtless folks, especially some of our present governments, convince themselves that because they are not feeling the stress of global warming and the changing climate personally, it isn't happening. Some even promulgate, from their own personal authority status, that it is all fake news, even the greatest conspiracy perpetrated on humans and koalas. However, as I write this humble explanation of what is happening with the biomes, the ecosystems, and the biodiversity that support our life, the world has again experienced the hottest year on record; heatwaves have taken their toll on biodiversity, wildfires are raging, ocean levels are rising, and the culprit is global warming. We can only nervously wait to see what next summer will bring regarding bushfires given that meteorologists have had to add a new colour to their weather charts and maps to illustrate the increase in temperature.

For my part, I am not a climate scientist. I am an ordinary person who has been following the climate science for forty years. I am now seventy-three, but thirty years ago, I thought that I would not live long enough to see the scientific predictions of what was to become of us come to fruition. However, I am still here, and yes, I am witness to those many predictions about the impact of global warming because of our activities, coming to pass in real time. As I stated earlier, I used to sit on my deck at home and think, *I cannot feel climate change, and I cannot see the changing climate.* Life was just perfect for me; so could the predictions really be true? Could it really get to be so bad for my children, who would have to survive global warming generating a changing climate and the degradation of ecosystems that are vital for life? I no longer think of my children and their possible demise because they have joined me as adults and are engaged in 'business-as-usual'. I now think of my beautiful, innocent grandchildren, who will bear the brunt of the impending environmental disasters through no fault of their own. When some not-so-well-informed people say to me that our grandchildren will be all right, that they will adjust to their new way of life, I do not feel that confidence because their lives will be completely out of their control. Nature will be in control then. My grandchildren are part of Earth's biodiversity, which has evolved to live in a stable biosphere and climate, and only those creatures that can adapt will survive.

In spite of dozens of international conferences, summits, and commitments to bring carbon emissions down, thousands of proactive environmental organisations pleading for action, heads of countries and states declaring that we are in an environmental emergency, and promises to cut toxic global warming pollutants in the atmosphere, nothing of consequence has happened on any scale that is effective. Earth is still warming and will continue to warm. Today life is 'business as usual' for most of us, even though evidence of environmental degradation because of a fast-changing world is clear for all to see – that is, for those who want to see. It seems that no matter how vocal Mother Earth is in announcing that she has had enough of our behaviour and putting on grand performances to prove her point, we still have not registered that life as we have known it is in serious trouble, and time is of the essence to respond to her very clear messages.

CHAPTER TWO

Three Flashpoints for Alarm

As a teacher of environmental studies, I, like the climate scientists, have downplayed the seriousness of our plight. In the past, no matter what climate scientists concluded from their research, they nearly always said, 'But we still have time' to turn our downward spiral around – not so today. Some are clearly saying that 'time is almost up'. Their only uncertainty is what is called 'tipping points' and the 'domino effect'. Chain reactions and feedback loops are clearly understood – that is, if this happens, then that will inevitably follow. Ecosystems of Earth are interlinked, so the big unknown is which link will give in first.

Climate scientists in their hundreds, if not thousands, agree that we have a meagre ten years at most to slow the worst of what is to come, so stay tuned. For myself, I have got off my safe, blissful deck and taken a good look at what is happening to our home planet. I have joined the dots, so I needed to put pen to paper as my commitment plan to do something more than 'business as usual'. This was my call to arms, so to speak. Climate scientists have put the whole world on notice that our current pathway to pursuing unlimited economic progress, unlimited economic growth at any cost to the health of Earth, is not sustainable, and therefore, it is not going to have a happy ending. Whatever we do to the other-than-human world, we do to ourselves.

We are called to undergo an ecological conversion. We have to understand at the deepest level that we are interrelated to all life, that we are interconnected with all life, and that we are totally interdependent on all life. Human life absolutely depends on the health of all the ecosystems and biodiversity that made life possible and continually supports life; life on Earth has been sustained because the balance and harmony of ecosystems has made it possible. It is us who are clearly upsetting the balance and harmony of the natural world. This is the scary part, so brace yourself for a layperson's interpretation of the science regarding what is going on in our world today. There are three major players causing our environmental emergency. They are global warming, which is responsible for the changing climate, pollutants in the atmosphere, and the loss of ecosystems and biodiversity.

Global Warming:

- As long as I can remember, climate scientists have been saying from their modelling that it was imperative to keep global warming under two degrees Celsius, but now they are saying loud and clear that it must be kept below one and a half degrees Celsius, and why is that? It is because the impact of rising global warming/heating is already visible for the eye to see. The icecaps are melting, glaciers are disappearing and receding at an unprecedented rate, oceans are warming and rising, and droughts are increasing in duration. Floods, fires, and famine are tell-tale signs as they have increased in intensity – no scope for 'I don't believe that the climate is changing'! We cannot go on kidding ourselves that it is not real as eighteen of the nineteen hottest years ever recorded have occurred since

2000, and 2021 is on a trajectory to be the hottest year on record. How hot does it have to get before we take notice? The global temperature in the ocean and on land has already increased by 1.2 degrees Celsius. Of course, there is an inbuilt discrepancy in the modelling depending on whether we urgently address the issues causing global warming or, alternatively, Mother Nature trips a 'tipping point' and brings on a 'domino effect'.

We need to stop talking about climate change as though it is futuristic. Saying that we have got to 2050 does not help to get zero emissions on track; 2050 would be at least seventy years since we have been told by climate scientists that we needed to cut greenhouse emissions. The projection is that 2050 will not be a favourable experience because there is every chance that we will have raised the global temperature above two degrees Celsius if we do not act decisively now. The terminology should be that global warming is happening, and we need to prepare as best we can to mitigate the worst outcomes of it. A consensus among 98 per cent of climate scientists is that we will get to two-plus degrees Celsius and possibly higher as the wheels are already in motion and cannot be stopped without drastic action now. A rise in temperature of two degrees Celsius could mean days with heatwaves upward to fifty-plus degrees Celsius; air conditioning will fail; and old and young will die. Plants, animals, and insects will be lost to extinction. The projection is that by the turn of the century, we could get to three or four degrees Celsius of global warming. Well, who cares about that statistic? It is totally irrelevant to action today as we may not be here as a species to endure it. There is ample scientific literature on the possibility of extinction of the human species; that is something we do not want to contemplate. Most of the natural world that supports us will be seriously disabled or wiped out as every creature has evolved to live in specific climatic conditions, and the human species is largely water, so why do we even think we can survive such catastrophic heat or a frozen Earth, for that matter? It is undeniable that even if we stopped greenhouse emissions going into the atmosphere today, the global temperature would continue to rise as there appears to be a delayed action or consequence; that is why resilience, mitigation, and adaptation are vitally important now – so that we can endure the changing climate until it restabilises. Each climate science report is another clarion wakeup call that we must drastically cut our greenhouse emissions.

Pollutants in the Atmosphere:

- A decade ago, I joined an environmental organisation, 350.org, that was doing its best to keep pollutants under 350 parts per million (ppm) in the atmosphere. This was deemed the last safe level that could sustain life in perpetuity. At the beginning of the Industrial Revolution, parts per million in the atmosphere were 280. That was a healthy amount for life on Earth. Pollutants in the atmosphere have increased dramatically with the Industrial and Technological Revolutions to measure today as 410+ ppm in the atmosphere and will continue to rise. So why is that significant to note? The pollutants in the atmosphere brought about by burning oil, coal, gas, plastics, and forests are the root causes of global warming. Global warming is driving the changing climate. No faith required here – it is measurable, even visible and tangible in some countries, and the consequences of current global warming are self-evident. Charts and graphs clearly show the correlation between human activity increasing greenhouse gases in the atmosphere and the rise in global warming. Meteorological instruments do not lie.

Loss of Ecosystems and Biodiversity:

- Every ecosystem on the planet is currently stressed and showing signs that the pressure humans are placing on them is getting beyond their ability to keep Earth in the state of balance and harmony that is so necessary to continue to support life. The current loss of biodiversity does not bear thinking about as over the last thirty years, approximately 50 per cent of biodiversity has been lost to the Earth community. Many creatures have become extinct or are threatened with extinction because of global warming changing the climate, the loss of habitats, and depleted food security. The records of loss of species are already well documented, and the statistics are truly alarming. It is now well documented that many species are going the way of the dodo on our watch. If Earth is going to restore its balance and harmony, the Earth community cannot lose any more of its biodiversity.

In summary, global warming has not been stopped, it is not slowing down, and it is not levelling out; in fact, it is getting worse and accelerating faster than predicted, especially at the icecaps. You might like to check it out and ask yourself, 'How does that affect me as an earthling?' Remember, we are all interlinked – no escaping that fact. Greenhouse gases are still increasing in the atmosphere and will continue to increase while the Earth community of people continue to burn coal, oil, and gas into the atmosphere. Added to the main culprits are the depletion of native grass vegetation and forests vital to drawing down carbon from the atmosphere. Although we are *Homo sapiens* – that is, wise humans – we do not know the full implications of the loss of biodiversity on the balance and harmony of the living Earth. What we do know is that the combination of the three drivers of climate change and environmental degradation are currently in free fall, and time is of the essence to address all three to stem the tide of an environmental catastrophe tending towards environmental collapse. After this tirade, my intuition is to say, 'But don't fret. All will be well' – but no, not this time. All may not be well.

This book offers a simple and brief introduction to many of the environmental issues to be concerned about. There are dozens of scientific papers and books written on each of the topics as well as YouTube presentations which you can follow up on if some area of interest captures your imagination. It is impossible to be on top of everything environmental as changes to the ecosphere are happening at rapid speed now. Every year statistics on the health of Earth are being recalibrated to reassess the ever-changing climate, which impacts on all life in some way.

CHAPTER THREE

Demystifying Our Environmental History

We have a timeline that shows how quickly in the human story we have come to upsetting the balance and harmony of our home planet. If you are under forty years of age, then you may not know that you were born into an environmental crisis. When you were born, the Technological Revolution had begun, and the mantra was, in most First World countries, 'Progress at any cost'. What was not realised was that the much sought-after 'progress' has come at a massive and continuing cost to the health of Earth. Speaking as a First World person, the problem is that if everyone on the planet lived like me, then we would need five or six Earths to sustain us; this, of course, is unsustainable as Earth is a closed system. Earth has all the gifts people need to live, but some of them are finite; once they are used up, their natural state has changed, and they are gone forever.

Mystery of Life — Four and a Half Billion Years Ago

To put the story of our environmental undoing into perspective, it is valuable to look closely at the story of us and the Earth's story. Let's start at the very beginning of life; we will call this reflection the 'mystery of life'. The origin of life on Earth is still somewhat of a mystery as physicists or cosmologists still cannot explain what caused the Big Bang. Scientists are still pondering on why, where, and how life began. As far as we know, Planet Earth – our home, nestled in the Milky Way Galaxy – is, to this day, alone in the universe as a planet that enabled life to begin and continually support life over many, many millions of years. Earth is one planet amongst billions, and scientists and scientific probes are continually searching the universe for evidence of life; to this day, Earth is the only planet to have life on it and has amazing ecosystems to support life. So what value do we put on the mystery of life when we are part of it? Do we value it enough to preserve the immense diversity of life that we are blessed with – that is, the totality of biodiversity that enriches our lives?

What do we know about this life-giving planet we call home? Planet Earth is strategically situated in relation to the sun. It is not so close that the temperature is too hot for life and not so far away that it is too cold, so this is sometimes referred to as a 'Goldilocks' position. Earth itself had a tumultuous beginning with being bombarded with debris from the Big Bang for over ten billion years. During this time, it was a fiery furnace until it finally cooled down. The centre of Earth is still a fiery furnace and reminds us of this with volcanic activity as molten lava rises up through the crust of Earth from time to time.

It is difficult to pinpoint when life began, but some scientists believe it could be as long as four billion years ago. The evidence for this is found in Australian coastal water in the form of stromatolites still visible today off the Western Australian coast. Scientists are still unravelling the mystery of life, and there are a number of theories proposed to explain its existence. For example, did life simply evolve from nothing, or did the seeds of life arrive by meteorites or comets crashing into Earth from outer space? Water is the key to life on Earth, and it is still a great unknown as to how water arrived on Earth given that the surface of Earth is two-thirds water. How life emerged on Earth is still something of an enigma and perhaps a miracle because so many conditions had to come together to enable life in the first place.

On a daily basis, we do not think about the fact that we live on a planet that hangs like a beautiful jewel in space, but we should. The powerful and majestic image that astronauts photographed as they rounded the moon of Earth hanging in space changed the way we perceive our home planet; we are earthlings, people of Planet Earth. Sometimes Earth is referred to as a spaceship because everything we need is on this planet. It is also referred to as a 'closed system' because nothing is lost or added to Earth except stardust that still lands on Earth. Everyone and everything we love is on this planet, so the most important response required from us in thinking about the mystery of life is to learn to love our common home, Planet Earth, even more than we love the country we were born into. People are prepared to die for their country, so are we prepared to do everything but die for the love of our home planet?

From the earliest evidence of life, from fossilised cyanobacteria evidenced in stromatolites, we know life began its awesome journey through an evolutionary process. From simple organisms to complex creatures, life emerged resplendent with an incredible multitude of biodiversity. The theory of evolution helps us to understand the complexity and diversity of life. As life exploded into being in the ocean and on land, adaptation was the key to 'survival of the fittest'. 'Adapt or die' was the order of the day for plants and every other creature. How plants and other creatures adapted to different biomes – that is, climate, terrain, vegetation, and food security – is an amazing study in itself. As we understand more about the lives of creatures, our minds are full of wonder at their ingenuity to survive. Just think about insects and birds, for example; they experience the same climatic and weather conditions as us, but they do not have the protections that we humans have been able to build to protect ourselves. As members of the human species, we have our own extraordinary story of evolution. Understanding something of the mystery of life helps us to appreciate our place in the cosmic, Earth, and human stories. To understand how we came to be in an environmental crisis, it helps if we fully understand our historical evolutionary story as a species so that we feel compelled to commit ourselves to deep adaptation and deep ecology to preserve all life on Earth.

Hunters and Gatherers — Two Hundred Thousand Years Ago

Fast-forward about four billion years to the first people, the evolution of the human species in Africa. According to anthropology and palaeontology, us black, brown, white, pink, yellow, or red people are descendants of hunters and gatherers who emerged from Africa, which gives us a common ancestry. For the longest period in human history, people hunted and gathered their food. The men were usually (but not always) the hunters, and the womenfolk were engaged in gathering food. Two words that are synonymous with hunting and gathering are 'foraging' and 'nomadic'. The people collected food from wild plants and animals, and when their food supply was depleted in an area, the hunting and gathering people moved on. For over two hundred thousand years, hunters and gatherers lived on, moved from, and returned to their land after giving plants and animals time to recover.

For thousands of years, people in family groups or tribes roamed from place to place, each time only taking what they could carry. They lived very lightly on the Earth and left very little behind to tell of their passing. With no refrigeration, hunters and gatherers only took from the Earth by way of plants and animals what they needed for their daily survival as they could not store food for lean times. Nothing was wasted; everything was used. Shelters were made out of rocks or brush, or where available, caves were used as domiciles.

Still today, there are people who are hunters and gatherers, such as some indigenous people of Australia and others who live in forests or deserts – that is, land that is not suitable for agriculture. They continue to live the life of their ancestors – that is, people who have adapted to climate, terrain, and food security. Some may think that their lives are difficult, but people who still live a nomadic and foraging life are, according to anthropologists, very content with their lifestyle and also very healthy. Their footprint on Earth was so light, it can barely be measured; there is so little evidence of their two hundred thousand years as nomadic people.

For hunters and gatherers, Earth literacy was and still is very important. They know their land intimately and all the creatures that share their land. They have a profound respect for all life, plant or animal, and much of their spirituality focuses on the Earth and its gifts. Earth wisdom was handed on from one generation to the next, usually through storytelling of the mysteries of life. Hunters and gatherers were environmentally literate; they could read the signs of the Earth and the creatures around them. Without perhaps expressing it, they knew they were deeply connected to the natural world and absolutely dependent on the natural world. Their survival depended on knowing what each ecosystem contributed to their lives and the importance of conserving life-sustaining biodiversity.

Through social evolution, over thousands of generations, nomadic life became more complex as they began to understand plants. Rudimentary horticulture was practiced as they observed that some plants were more beneficial to them than others, so some seeds were strategically left to flourish, whilst other less useful plants were cleared. These skills served them well until a major social and environmental transformation took place, known as the Agricultural Revolution. Unbeknown to the nomadic people, this was the beginning of human demise, our loss of intimate connectedness to the land and respect for the natural world.

Agricultural Revolution — Ten Thousand Years Ago

About ten thousand years ago, over time, a transitional change took place in the way people lived. The Agricultural Revolution changed forever the relationship humans had enjoyed with the land. Agriculture – the growing of crops, along with the domestication of animals – provided a new way of life for once nomadic people who discovered that there were certain types of plants in the wild that produced seeds that they could collect and plant in abundance, cultivate, harvest, and store for food to support them throughout the entire year. Food security was assured, and so people began to abandon their nomadic lives for a more settled lifestyle.

Because this period of the Agricultural Revolution preceded writing, not a lot can be known with absolute certainty when humans made the transition from hunting and gathering to a more sedentary lifestyle in settled villages. It is possible to imagine that it would have happened over time, in different locations, and that there were a number of factors that influenced the change. As wisdom and skills to support permanent agriculture developed, food security was assured, and so villages became the norm, larger populations gathered together in more permanent housing, and people enjoyed a more social existence. To ensure their survival, villages tended to be established near a permanent water source, along rivers or around lakes.

The Agricultural Revolution was underway first around the Fertile Crescent in the Middle East, Egypt, and India as well as, independently but simultaneously, in China. These people were among the first to plant, sow, and harvest in season as the climate allowed. Understanding climate and seasons, the soil and rainfall would have been paramount to them for the successful growing of crops that would provide them with food security.

From an environmental perspective, the Agricultural Revolution was the beginning of change in our relationship with the land and animals. It is scientifically recognised that nomadic people cultivated the land by burning the grass before they moved on to encourage new growth and return animals to the area, but this practice did not have a lasting impact on the health of the land. It was the Agricultural Revolution that really began dominance over the land as trees were cleared for land for cropping and heating or used to build more permanent housing. As their agricultural skills developed, they realised that they could plant more crops than needed, and so they were able to produce food to trade with other people. This was the beginning of enterprise marketing, which opened up communication with other villages.

As people conquered the land, they began to lose some of their close connection with the natural world. Instead of working with the land in its seasons, people began to take control of the land, vegetation, water, and soil. With population growth and people to be fed, land was needed for agriculture, and so people began requisitioning land from native forests and grasslands. The Agricultural Revolution commenced human management of land and animals, and it continues today on a grand scale; the reality is no land is safe from human interference and domination.

Industrial Revolution — Two hundred and fifty Years Ago

The next big event in environmental history was the Industrial Revolution, which began about two hundred years ago. The

Industrial Revolution spearheaded the amazing life we enjoy today. It was a glorious age of inventions, creativity, and human progress. It heralded an era of mass production and a consumer-based economy. There seemed to be no limit to the improvement of life for many. It was the age of mechanisation; machinery of all types was invented and manufactured to replace manpower and horsepower.

The powered engine to support the Industrial Revolution came initially from burning wood and coal, followed by gas and gasoline from oil. Factories were built to house machinery and workers, and it followed that villages became towns and later cities. The Industrial Revolution meant that workers were in great demand, and many who had worked the land sought a new life for security in the growing cities. Farm machinery initially pulled by horsepower supported and increased the Industrial Revolution. It was a time of great progress, and the future was full of promise. Factories required increasing amounts of raw materials such as timber, cotton, and steel, and more land was cleared of trees for agriculture, fuel, and buildings. Every raw material was required to advance the Industrial Revolution and the golden age of prosperity. It was the beginning of consumerism and a disposable society, and because things were mass-produced, everything could be reproduced. There was always a bigger and better, must-have, exciting new invention or commodity.

This boom of supposed progress initiated new environmental challenges such as pollution. Smog from factories and the burning of wood and coal filled the air. Waste materials from factories and untreated sewage found their way into water systems, with serious repercussions for the health of rivers. Waste was a new phenomenon, and landfills were created to cope with ever-increasing amounts of used and unwanted items. Deforestation, air pollution from burning fossil fuels, and waste are the horrendous downsides of the era of insatiable need for progress at any cost. The wonderland that nomadic people had enjoyed was seriously threatened by the beginning of the degradation of Earth's ecosystems vital to life.

At the end of World War II, there were about three billion people; today there are more than seven billion people, with the prediction of rising to eight or nine billion before levelling off. Everyone has to be housed, transported, fed, and refreshed with life-giving water from the gifts of Earth. Everything we have comes to us as a gift of Earth, and Earth's gifts are not simply resources to be used and abused but rather to be valued and used again and again. The industrial thinking was 'cradle to grave', but today's thinking must be 'cradle to cradle' – that is, almost everything must be able to be remade into something useful.

The Industrial Revolution generated an exciting era in human 'progress', but it came at a huge cost to the environmental health of Earth. The thinking clearly was that the gifts of Earth were there for the taking, to be plundered and exploited without conscience or consequence. A further unravelling of the health of Earth was to come at rapid speed – that is, within the lifetime of older people today. Our generation, over the last fifty years, has used up more of Earth's gifts than all previous generations combined. That is a very salutary thought and one worth pondering on given that many of Earth's resources/gifts are finite.

Technological Revolution — Sixty Years Ago

The Technological Revolution is a recent phenomenon in human and environmental history. To put the Technological Revolution into perspective, about sixty years ago, there were very few technologies – for example, no plastic stuff, flying around the world was not an option for most people, fewer televisions, no mobile phones, no personal computers, and, worldwide, very few people owning cars. Technology has invented a whole new way of being in the world. It has been an extremely creative period in human history as humans have invented the most mind-boggling technologies and appliances for every purpose. For those who are living through the Technological Era, there is nothing to compare the change, over time, as to how human society operates. The magnificence and splendour of this era has no rival; it is a time of incredible creativity and change, but it has all come at a gigantic cost to the health of our home planet. It is the Technozoic Era, the era of complete human domination of the resources of Earth.

A definition for the Technological Revolution is an era of increased control of the environment. From manpower to horsepower to fossil fuel power – that is, energy captured from the sun and stored in coal and oil – there are no limits to 'progress at any cost'. Human domination of the natural environment is technologically and physically complete. No ecosystem is safe from human activity, and every ecosystem is viewed as a possible 'resource' for further 'economic progress' to power the Technological Revolution. However, everything we have is a 'gift' of Earth, and gift giving is supposed to be reciprocal as well as requiring gratitude. For example, technology has given us the internet, a virtual highway to knowledge; we cannot imagine life without computers, so complete is our adaptation to a modern way of communicating with one another. With the advent of information technology, news and views provide instant gratification. The Technological Revolution has also conquered distance with the introduction of many inventions and innovations for how we get around. However, the material for every invention comes to us as a gift of Earth.

To engage in ecological conversion to address the environmental crisis means that we have to get our heads around the fact that the benefits of the Technological Revolution have come at a huge cost to Earth that is not sustainable in perpetuity. Raw materials – that is, oil, coal, timber, agriculture, and produce of all kinds – are in great demand and are being shipped around the world at a most alarming rate. This process is called globalisation, servicing a global market, a two-way exchange of goods and services. However, as we extract more from the Earth for the sake of the economy, we will be left with a depleted Earth as many gifts of the Earth are finite. We may not understand how finite some gifts are given that we do not comprehend the full impact of global warming on the extinction of species such as insects or fish.

The worst outcome of the Technological Revolution was the initiation of a consumerist and disposable society. Waste is at an all-time high; we put our bins out to be picked up by the garbage contractors, but where does our rubbish go? In environmental terms, there is no such place as 'away'. 'Away' is somewhere, and our waste is there in that landfill. There is always a new gadget or a new product to purchase; then it is out with the old and in with the new.

The Technological Revolution has also dramatically exacerbated the pollution of air, water, soil, and the ocean to the detriment of wildlife on land and in the sea. Many of Earth's gifts are finite, and we are just realising that our wonderful lifestyle cannot be sustained without changing our relationship to nature. Planet Earth and its gifts are all we have forever, and right now we are using Planet Earth like a credit bankcard that will give us grief if we cannot pay it back. We keep drawing on Earth's capital, but there is no way we can pay the Earth back once we have spent the resources.

Beginning with the Agricultural Revolution, moving through the Industrial Revolution, and now experiencing the Technological Revolution, the changes in human society are registered on an environmental time clock – that is, how these changes and the way we live have impacted on the health of Earth. New phenomena such as an insatiable need for raw materials, unimaginable pollution on a grand scale affecting every ecosystem, thoughtless waste, and population growth are key factors in advancing the environmental crisis. Now we are in an environmental emergency tending towards an environmental catastrophe.

It is so very important that we learn about the health of Planet Earth. Earth literacy is a 'must learn' for everyone. We are challenged to think differently about the Earth – that is, Planet Earth is not an *infinite resource* but has *finite gifts*. If we can change the terminology we use to speak about the Earth and its resources/gifts, then we are in with a chance to change our hearts; if we change our hearts, then we will change how we live and vote. We must be part of the solution because we are the environmentally educated people of Earth. Planet Earth needs everyone to be reflective and creative to devise new and amazing ways of being in the right relationship with the natural world. Why is that? The human species, over the last fifty years, through our multiple activities, has quite clearly changed the chemistry of the air, the ocean, and the soil. We have changed the way the Earth has functioned for millennia.

Sadly, we have done a great deal of damage to Earth. Our relationship with nature needs some heavy-duty counselling to fix it – and fast. Falsely, humans think we are the dominant species, that we can do whatever we like to Earth without conscience or consequence. If we recall the conditions that allowed life on Earth to evolve and flourish, then it has to be accepted that we are serial destroyers of those life-giving conditions. It is strange how people keep talking about 'saving the Earth'; actually, the Earth does not need saving. Planet Earth was doing just great before we evolved, and if we were to self-destruct because we failed to respond to Mother Earth's messages, then Planet Earth would thrive once again. There are many creatures that, no doubt, would like to see our backs. We are at war with the rest of creation; sadly, at this point, we are winning, but the tide is turning, and nature is pushing back with unprecedented fire, floods, and famine. Still, it is important to remember that Earth has an amazing ability to renew itself given half a chance. The lockdown of people during COVID-19 has proved that with us needing to isolate at home, nature took a refreshing breather for a while. Environmental literacy, if it really woke us up, could mean our future is full of promise and very exciting. We are all part of an environmental education revolution, another era in human history.

Environmental Crisis

We are currently in an environmental crisis of our own making, and we have come to this point partly because of our ignorance but more realistically our arrogance. In our pursuit of a better life, of increasing our prosperity and desire to fulfil all our needs and wants, we have engaged in 'business as usual' without a thought of what we are doing to the natural world that supports us. After ten thousand years of the human species living a sustainable existence, we are now living an unsustainable existence. It is most fortunate that we know how we came to be in an environmental crisis because we also know how we can reverse it. We know what we have to do, and that is to realign ourselves with the other-than-human natural world. We have to relearn how to live and work with nature, not continually against it. A species can only thrive if everything around it thrives as well. The world and our lifestyles are of our own making, so we are in with a chance to remake our world and our lifestyles to be cognisant and sensitive to nature.

Every week documentation is being produced to alert us to the impact that our lives are having on the natural world, and it is usually a 'gloom and doom' situation if we continue along our current pathway of environmental injustice. Statistics and photographic evidence provide the data for a 'blow by blow' commentary on the state of the natural world, which is normally not a good news story. There are a number of factors that have brought us to what many are calling an environmental emergency. These factors include global warming, pollution, waste, urbanisation, population growth, overconsumption, insatiable economic growth, the limitless need for raw materials, affluence, poverty, and conflict, which, in turn, have resulted in climate change, environmental degradation, and the loss of biodiversity on a grand scale. We have not stopped to understand and appreciate that we cannot go on consuming the Earth's gifts without thought of the consequences. We have been mentally fixated on the notion that more is better, that continual progress is possible, that we can continue to dominate nature to do our bidding, and that we can continue our assault on the Earth's resources to the point where we have changed the topography of the Earth. Sadly, for us, there will be a day of reckoning with Mother Nature.

Some readers may have seen the film *Independence Day*, where aliens come to Earth to take it over. It was said by way of explanation for their invasion that they fly around the universe seeking planets to colonise, devour the gifts of the planet, and move on to another planet. There is something sinister, but familiar, about this story as humans are devouring Earth's gifts and have a space travel vision to colonise other planets or the moon. What are we thinking? Earth is the planet we evolved to live on. If the wealth of the consumerist world can be measured by an escalation of consumerist goods, then the downside is that the Earth's gifts are correspondingly and continually being depleted at a comparable rate to provide those consumerist goods, and of course, it follows that it is unsustainable and that, ultimately, it is to humankind's disadvantage. Therefore, the root causes require attention rather than simply mitigating the philosophy of infinite progress. Scientists are trying to fix problems with Band-Aid solutions – such as heat-resistant trees or harvesting coral eggs and sperm to unite them with other coral eggs and sperm that may cope better in a warming ocean – rather than stop the problem at its source, all because we are not listening to them, and they need to do something to preserve life on Earth.

In summary, we now know how life began on Earth, even if there is still some mystery around exactly how life came to Earth to evolve. We know that the human species evolved around two hundred thousand years ago, and for nearly all that time, they were nomadic hunters and gatherers who walked lightly on the Earth. Then over a very short period in human history, massive and irreversible change took place because of the Agricultural Revolution about ten thousand years ago, which began domination of the land, followed by the Industrial Revolution about two hundred years ago, which initiated pollution and waste, followed by the Technological Revolution's insatiable need for raw materials, which has exacerbated pollution and waste because of a huge paradigm shift to a consumerist and disposable society. So we know how we came to be in an environmental crisis that is now giving climate scientists so much grief regarding the continuation of the human species and all other creatures as well. All creatures evolved, along with the human species, to live in specific climatic conditions, and sadly, those conditions are being changed by our inconsiderate and perhaps thoughtless selves. We are the perpetrators causing an environmental crisis, an environmental emergency that may well lead to an environmental catastrophe or, worse, environmental chaos if we do not take our survival plight seriously now. There is no more time; it is the eleventh hour and ticking awfully close to midnight.

CHAPTER FOUR

What, in a Nutshell, Is Happening to the Biomes?

A study of biomes is to look at a panoramic picture of the complexity of the biosphere. The Earth is divided into natural areas called biomes. Which biome is your backyard in? There are different classifications of biomes, but this reflection will present five for consideration: tundra, taiga, forests, deserts, and grasslands. In simple terms, biomes are broadly distinguished by their distance from the equator, both sides of the equator, but they do merge in different countries. They are generally defined by their climate, rainfall, topography, and geography as well as the plants and animals that live there, and within each biome, there are many varied ecosystems. The human species lives in every biome, and generally, humans are the dominant species who have polluted, destroyed, or exploited the gifts of each biome, so it is valuable to understand the contribution each biome makes to the living nature-scapes of Earth. Until we appreciate what each biome contributes to the balance and harmony of a healthy planet, we will not feel compelled to be a voice for the voiceless biosphere.

Tundra Biome

The tundra biome is one we do not think about very much in Australia because it seems so far away for most of us, but we are one Earth, and what goes on in the tundra affects us in some way. The tundra biome is located in the latitudes just below the icecaps of the Arctic ecosystem. It extends across North America, much of Alaska, a good deal of Canada, the top of Europe, and Siberia in Russia.

Tundra actually means 'a land devoid of trees', 'a treeless plain'. It is one of the coldest ecosystems on Earth and experiences very low rainfall. The tundra is in the Northern Hemisphere, it circles around the Arctic Circle, and it covers about a fifth of the Earth's land mass, so it is a very big biome, and a number of countries experience life on the tundra. The temperature can drop to minus thirty-four degrees Celsius in the dead of winter, but the temperature can rise to around twelve degrees Celsius in the summer. The tundra experiences very long winters and very short summers, and as a consequence, very little grows there because plants need sun for energy to grow and thrive; however, there are about sixty days of sunshine a year. Many plants have adapted to the cold and long winters, and because plants live there, so are there insects, birds, and animals that also call the tundra home. Some creatures hibernate through the worst of the long winter chill, while other creatures opt to migrate to warmer climates.

When we think 'tundra', we think permafrost – that is, permanent frost. Permafrost is that top layer of soggy ground when snow and frozen rain begin to thaw in the very brief summer months, although there is also a sub-layer of permafrost that remains frozen all year round. Permafrost is a very fragile ecosystem that is beginning to unravel today. Scientists use the tundra to research the Earth's past because it is frozen and unchanged by humans. They also research what is happening to the tundra as the Earth warms up because of human activity. As the permafrost thaws because of global heating, too much carbon dioxide and methane are being released into the atmosphere. The tundra is called a 'carbon dioxide sink' as it stores decayed plants and animals under its ice.

Traditionally, not many people lived on the tundra because it is so cold and wet underfoot. However, exploration for oil has brought more people, and towns have grown up to service the oil industry. This is giving environmental scientists deep concern because the biodiversity – that is, the plants, insects, birds, and animals that live on the tundra – are delicately balanced, and human interference can seriously challenge the safety of the creatures that live there. Habitats, food security, and migration routes are at risk of being disturbed to the detriment of native wildlife on the tundra.

The tundra provides a record of any change in temperature in the past and the present. The tundra could well be a 'canary in the coal mine', an early indicator of potential danger. Scientists are already recording huge changes in temperature in the Arctic Circle. In June 2020, days with temperatures of thirty eight degrees Celsius were reported. This rise in temperature will have a huge impact on melting the tundra permafrost, thus releasing more greenhouse gases.

The main concern is that if the tundra continues to melt, carbon dioxide and methane, captive under the ice and snow, will continue to leak into the atmosphere. It will influence the air we breathe as well as the changing climate around the world. More importantly, we need to understand the contribution the tundra makes to the health of the planet and, as much as possible, to leave the tundra in its natural state, where the people and creatures that live there can enjoy a peaceful life in their traditional habitats as well as enjoy food security.

An interesting piece of information is that as the tundra melts because of global warming, the remains of mammoths that were victims of the last Ice Age are being found. Mammoth tusks are huge; some are over six feet or two metres in length. While tusk fossickers are delighted with their finds as they are very valuable monetary wise, the fact that many more tusks are being randomly found over the last few years is a cause for concern because the ice and permafrost must be melting to reveal them! Climate history is in the revealing of mammoth tusks, but also, the fact that they are being found in numbers indicates an environmental permafrost meltdown in the tundra because of the warming climate today.

What, in a nutshell, is happening in the tundra ecology?

- There is a lot going on environmentally on the tundra biome.
- The tundra is warming at an alarming rate – that is, much faster than the rest of the world. The melting of permafrost will supercharge global warming.
- Alarm bells are ringing because the tundra is experiencing persistent warm weather, increasing wildfires, which causes permafrost to thaw quicker.
- Thawing permafrost releases large amounts of methane, a very potent greenhouse gas which is twenty-eight times or more the impact of carbon dioxide in polluting the atmosphere.
- The released methane then goes into the atmosphere mix and circulates around the world. We are all subject to the consequences of thawing permafrost.
- Siberia – which is usually cold, icy, and wet – is thawing and setting up conditions for wildfires. This is a rapid change, and the thinking is that wildfires could become the norm in Siberia and across the tundra.
- The tundra has some very clear feedback loops: melting permafrost increases pollutants in the atmosphere, which

increases the world temperature, adds to global warming, and initiates the thawing of more permafrost, which then releases more heat-trapping gases, dries the soil, increases wildfires, and adds to global warming.
- The thawing of permafrost is considered by climate scientists to be one of the 'tipping points' that can influence serious and irreversible changes to how Earth's biomes function when exacerbated by the changing climate.

The thawing permafrost has got climate scientists in a real dither. What is going on there is unprecedented and somewhat not anticipated in their climate modelling because it is thawing much faster than they thought it would. However, today the tundra is presenting climate scientists with a whole new scenario of potential catastrophic conditions. What is happening to permafrost is now captured on camera, such as houses sinking or collapsing because their ice foundations are thawing and subsequently damaging or exposing infrastructure. The thawing permafrost is eroding riverbanks, revealing the depth of permafrost.

As huge numbers of mammoth and other animal bones are being exposed, there is no denying that the problem is real. As global temperatures rise, the carbon and methane captured in permafrost could translate to serious planetary heating emissions within feedback loops. It is one to watch as it is changing rapidly because of unprecedented warming which shows no signs of slowing down. The tundra biome is a first-class example of the need to join the dots for understanding the bigger picture beyond our backyards. What is happening to the tundra biome will impact on all of us.

Grasslands Biome

Today more than 50 per cent of people live in cities, so grasslands are not something we think about, and they hardly get a mention in the media. However, they are critical ecosystems because our food is largely dependent on them. Over decades now, grasslands have been subjected to some harsh farming practices, especially the use of chemicals, to ensure a successful crop. Chemicals such as pesticides and herbicides as well as – pre-emergent, emergent, and post-emergent chemicals – are used as fertilisers, which is the norm in some countries. They are poured onto plants and soil by farmers who have been seduced into thinking that chemicals are a necessity to grow agriculture crops. It is, to some extent, a great unknown, the magnitude of damage that chemicals do to subterranean mini creatures – such as bacteria, fungi, nematodes, protozoa, and anthropoids – that live in the soil that are damaged by runoff chemical pollution as well as the loss of native grasses. Today many millions of wildlife creatures compete with humans for a safe habitat and food security on the grasslands biomes. Learning the importance and value of grasslands is therefore a sacred duty for all of us.

Grasslands biomes, in their heyday, covered about 25 per cent of Earth. Grasslands are called various names in countries that have extensive grasslands, such as prairies, steppes, pampas, savannahs, and rangelands. There are two major types of grasslands: tropical grasslands and temperate grasslands. Grasslands are generally located in terrain between deserts that are too dry to grow agriculture crops or pasture cattle and forests that usually experience more precipitation. Therefore, as the name suggests, native grasses were the dominant vegetation interspersed with very few trees. Not surprisingly, grasslands occur where soil, rainfall, and climate suit the growth of grass. Many millions of people and a myriad of creatures are at home on the grasslands biomes; for example, in America, the prairies supported millions of free-range buffalo that lived on native grasses, but there is no going back to free-range pasturing because of farms and fences. In Africa, the steppes still support animals such as wildebeest and millions of other animals – but for how long? Can you name an animal that is not being poached for human gain?

Over long periods, the grasslands biome provided habitat and food to many millions of plants and herbivorous creatures. Some of the largest creatures are herbivorous, such as zebras, along with smaller creatures such as kangaroos. In the great circle of life, herbivorous creatures are food for carnivorous creatures such as lions, which are at the top of a grassland's food chain, those amazing iconic creatures that make grasslands their habitat. Many of the wildlife inhabitants of grasslands are now threatened because grasslands are ideal for growing food and pasturing sheep or cattle for the human species. As the human population grows, so does the need to annex more grasslands for cropping and pasturing cattle. More cropping for human consumption means less grasslands for other native creatures such as elephants. As humans encroach more and more on grasslands biomes, many species of animals, birds, and insects are threatened with extinction. As each animal species is pronounced 'extinct', the domino effect will kick in, and all species that depend on grassland wildlife are at risk of extinction.

Grasslands are very suited to farming sheep and cattle. As people roamed far and wide across the continents, they took their livestock with them. They conquered grassland wildernesses and planted their monoculture seeds to produce so-called improved crops for sheep and cattle for human consumption. As a result, grasslands, as natural ecosystems or biomes, are seriously diminished. Scientists calculate that only about 5 per cent of native grasslands remain. In a little more than two hundred years, humans have dominated the grasslands biome. In some cases, cropping has been disastrous for traditional grasslands and for plants, insects, birds, and animals that have lived on them for millennia. Too late, farmers learned that not all grasslands are suitable for long-term cropping or grazing; for some farmers, their experiment left arid soils in its wake. Fortunately, many traditional dwellers are collecting native seeds to plant and reclaim some of the grasslands lost to deserts. Replanting native seeds has been very successful where it has been tried, which is a welcome result. There are now a number of seed banks around the world that are storing native seeds. The seed banks are intended to store native grass seeds in case they are needed for the future in the event that modern monoculture or genetically engineered crops fail.

Even though grasslands, in general, have fertile soil and sufficient rainfall, poor farming practices have occurred in some places, so soils have been denuded of nutrients so vital to plant life. Poor agriculture practices can ruin soil and turn flourishing grasslands into lifeless, barren land. In such cases, the topsoil can be easily blown away, with no rooted plants to hold the soil, thus reducing once thriving grasslands to desert-like conditions. The Dust Bowl saga in America in the 1930s is not only a good example of misusing fragile grasslands but also an example of reclamation.

What, in a nutshell, is happening to grasslands biomes?

- There is a great deal to be considered about what is happening to grasslands.
- About 95 per cent of native grasslands have been impacted on by the human species, with only about 5 per cent remaining as a natural habitat for wildlife.
- The human species is invading grasslands to the detriment of evolutionary creatures that call grasslands their home.
- This means the loss of an incredible amount of biodiversity so vital to keeping harmony and balance in the grasslands biome.

- The depletion of biodiversity by hunting native animals for their tusks and fur or just as trophies has occurred. In our arrogance or, indeed, ignorance, we do not know the benefit each creature brings to the health of a biome like the grasslands.
- After annexing native grasslands, some farmers turned to growing monoculture crops such as wheat or corn. Monoculture farming can deplete grassland soil of its nutrients, which weakens the biodiversity in the soil.
- Some grasslands have been subjected to agricultural practices that have already ruined the soil and stripped them of native grasses vital to their health.
- Where the cropping of grasslands has failed, the grasslands do not always return, but rather, the grasslands join the desert biome.
- When grasslands are converted to pasturing sheep and cattle, there is the danger of destroying native flora and, therefore, fauna that depend on the grasslands.
- Toxic pesticides poured onto crops are deadly to native flora and fauna.
- Grasslands are dependent on traditional rainfall, so with climate change, rainfall can change, and if less rain falls on grasslands, they can turn to deserts.
- Global warming could dry out grasslands that have been converted to crop and pasturing animals, leaving deserts in their wake. This will have huge implications for feeding the human population.
- About 70 per cent of crops grown on the grasslands go to feeding animals in feedlots and other animals for human consumption.
- Urbanisation is infringing on grasslands. Houses and infrastructure are being constructed on good, fertile grasslands that will be needed in the future to produce food.

- Habitats for native fauna are lost to invasive species when the balance and harmony of grasslands are interfered with.
- The traditional farming of grasslands, such as strategic burning, can be neglected.
- No biome is safe from the most dominant species on Earth.

Surviving on the grasslands biome is a challenge for the creatures that live there. As the climate conditions change and become more extreme, the grasslands' biodiversity of creatures will find it harder to survive. No creature on Earth is exempt from the effects of global warming, and therefore, the challenge is to adapt, migrate, or die. With diminishing native grasslands, the pressure is on the human species to rethink the wisdom of taking over the grasslands and establish farming methods that are more conducive to restoring as much of the native grasslands biome as is possible. There are plenty of examples in Western countries where farmers have turned to the intense irrigation of crops, only to raise groundwater, which, in turn, brings up salt so that the soil becomes saline, and nothing much will grow in salinity. The once fertile soil becomes a wasteland rather than a grassland. There are serious initiatives afoot by farmers to rethink their farming practices, such as regenerative farming, which used to be called rotation farming in earlier years. Free-range grazing as opposed to feedlots is now on the table for consideration, and if this idea is taken up, then perhaps it is possible to return some grasslands to their native state.

The precautionary principle must be applied to the grasslands biome by considering the long-term impacts of global warming on crops that have not evolved to survive extreme heat or unseasonal rainfall such as flooding or drought. It is imperative to restore as much of the grasslands biome by rethinking how human needs can be aligned with the needs of grasslands biomes to maintain balance and harmony for all biodiversity that live there.

Desert Biome

Deserts are formed by a weathering process brought on by wind, rain, and temperature acting on rocks, but even deserts will feel the brunt of the changing climate. Apart from the people who call desert biomes their home, not much thought in general is given to them. We are inclined to think of them as places that are not very inviting. We city slickers do not, in general, give deserts a thought, but they are uniquely fascinating places and worthy of our reflection.

Deserts are a natural phenomenon. Australia is largely a desert country, so most people who have migrated to Australia live in reasonably close proximity to the coast and into the hinterland to where the deserts begin. Deserts make up about 25 per cent of Earth's landscape, so that is a substantial area of land where very little vegetation can grow. It is calculated that half a billion people live in desert biomes and about 10 per cent of biodiversity call deserts home. They are sometimes called 'lizard' country as reptiles don't mind the desert because evolutionary processes have allowed them to adapt to the harsh environmental conditions.

Deserts are highly adapted biomes with natural ecosystems that provide services to the health of Earth as do other ecosystems. We who do not live in desert country don't think much about the plight of deserts; however, every biome is subject to the consequences of global warming, so it is not surprising that deserts have their story and fears for the future.

Hot deserts are usually located above or below the tropics of the equator and are very important and totally fascinating biomes or ecosystems. There are hot and cold deserts. An example of a cold desert is Antarctica, the largest desert on Earth. The Sahara in Africa is the largest hot desert, so what defines a desert is extreme heat or cold and little or no rain.

Not many creatures can survive such unforgiving conditions even after millions of years of evolution, so they are places that support very little biodiversity. However, there are still many creatures that have adapted to the extremely harsh conditions of deserts. Some insects and reptiles have adapted to live in the desert all year round, whereas many birds migrate across the desert. Birds still need to know where they can refresh themselves on their migrating journeys with water from springs – that is, water that flows upward from an underground reservoir known as an oasis. Creatures that make the hot deserts their home have to find water and shade somewhere because they cannot survive the harshness of the sun for lengthy periods; for example, ants emerge for only minutes at a time. No creature goes out in the noonday sun!

When we think of deserts, we may think of sand and endless sand dunes that are mobile as they shift according to the wind. Although deserts differ around the planet, they are known for extremely low rainfall. If it does happen to rain, it can evaporate before it hits the sand. Whatever little rainfall there is, plants have an amazing ability to absorb water and store it in their leaves and stems, such as cacti. The kangaroo is a good example of an arid desert creature that seeks shade for most of the day and feeds on desert plants in the evening and at sunrise before it gets too hot.

One of the most spectacular events in the Australian desert is the emergence of wildflowers. The seeds wait long periods in the ground until there is a huge downpour. Almost overnight, in some parts, desert flowers bloom, and the sight of so much colour in the desert is breathtaking. Other creatures can lay dormant for long periods but burst into life at the first sign of a downpour. At this time, the desert in Australia is fully alive. The desert in full bloom is a miracle that begets life.

An interesting phenomenon of deserts is that while they are extremely hot during the day, they can be cool to cold in the night as they cool quickly after sunset because they lack the humidity of the tropics. People travelling through deserts must prepare for cold nights, which the people who live in deserts know all too well.

Apart from natural deserts, desertification can occur through human activity where land is overgrazed by sheep or cattle or forests are cut down to provide land for cropping. When rain does not come, crops fail, and the topsoil can be blown away; in fact, millions of hectares of land are lost to desertification every year largely because of drought. Scientists have calculated that with increased drought conditions, global warming, deforestation, and the loss of grasslands, a third of Earth's land surface is threatened with desertification. However, in some countries, the replanting of forest trees has been very successful, and the practice of regrowing native species of grasses is continuing in different parts of the world, so that is a positive response to desertification caused by human activity.

What, in a nutshell, is happening to the desert biome?

- Desertification is an issue which means deserts are expanding as the climate and land use by the human species changes. Deserts are increasing all the time.
- Grasslands bordering deserts are always at risk.
- Deforestation has also resulted in land turning into deserts.
- Semi-dry deserts are becoming more arid and so less able to sustain life that has evolved to live there.
- Any change to deserts impacts on the diversity of life within the biome. It is thought that creatures that live in the desert could hold the key to survival in a warming planet.

- The temperature in some deserts is increasing; like the icecaps, they are getting hotter faster than other biomes.
- Warmer temperatures and fluctuating rainfall are presenting ever-increasing issues for desert dwellers.
- Climate change and human exploitation are impacting on the health of unique habitats and rare species in desert biomes.
- The human exploitation of underground aquifers is restricting desert life. This is an alarming issue for those who have researched and recorded underground artesian aquifers as some aquifers are not naturally replenished.
- A number of deserts are dependent on melt from glaciers, so if or when glacier melt is no longer present, then rivers through deserts will be dry, thus affecting desert life that depends on them, for example the Nile or the Colorado Rivers.

Australia is an ancient land with a large desert biome. Many indigenous people in Australia have lived and thrived in the desert for over forty thousand years, so even deserts are life-giving to those who respect it and live within its needs to remain healthy. Every biome is subject to human use and, if opportunity presents, exploitation. The scientists who study deserts argue that the desert biome has much to offer our understanding of how Earth functions as a planet. Again, it would be wonderful if people could experience desert life, even if it only meant that they gained a new respect for the biome they live in. Every biome enriches our lives, and we must not use and abuse it but rather love and protect it.

Rainforest Biome

The definition of a forest is land dominated by natural or native trees. Forests are distinguished from plantations that are, in general, devoid of native flora and fauna, such as palm oil or pine plantations. Forests, especially rainforests, are iconic creatures of the wild, and for hundreds of years, people have known of their important place in Earth's ecology. National parks have been established because some very intuitive, futuristic, and environmentally literate people wanted to preserve such magnificent trees and wildlife for posterity so that all generations can experience their awesomeness.

Plants and rainforests have a history of about four hundred million years. The evolution of tiny plants to forests has enabled trees and biodiversity that forests support adapt to changing climates and different terrains. Rainforest life has a harmony about it that borders on the mysterious. The service that a rainforest provides is reciprocal in its gift giving as every forest creature has an individual place and role to play. We tend to just think of the magnificent trees, but there is so much more going on in a rainforest, such as animals, birds, insects, reptiles, symbiotic roots, fungi, and a multitude of microorganisms at work to maintain a healthy rainforest.

Today forests cover over a quarter of Earth's land surface. There are three classifications of forests depending on their latitude in relation to the equator. They are tropical, temperate, and boreal. Tropical forests are closest to the equator. Boreal forests grow in the taiga biome, where forests are highly adapted to cope with an abundance of snow. The boreal forests have been defined as the largest terrestrial biome in the world, but we don't really appreciate its importance regarding the health of the biosphere. Temperate forests grow in the latitudes between the tropical and taiga biomes. The largest tropical rainforest is in the Amazon Basin in South America. The Amazon Rainforest has been referred to as the 'lungs of Earth' because of its awesome capacity to absorb toxins in the atmosphere and purify the air. However, over twenty per cent of the Amazon Rainforest has already been destroyed, and currently, the fear is that much more of the forest will be destroyed by logging to provide more land for agriculture. Deforestation of the Amazon is continuing at an unprecedented rate today. It is difficult to point the finger at other countries' deforestation practices when Australia is amongst the world leaders in chopping down trees. As a result of deforestation in Australia, we have a damning record on mammal extinction and threatened species. This is to our eternal shame, and future generations may not forgive us for not defending the trees of the forests.

The stupidity of destroying rainforests is that the topsoil is not always deep as many of the magnificent trees grow and are nourished from the humus decaying on the forest floor. In Kenya, where they chopped down much of their forests in favour of cropping, rain and wind rapidly eroded the soil and turned the once flourishing forests into desert-like conditions. The Green Belt Movement begun by Nobel winner Dr Wangari Matthai is desperately trying to reclaim the forests, so this movement is enjoying a success story well worth following. It is very important to appreciate rainforest ecology as it is a study not only of the trees but also of all the services forests provide plus biodiversity that make the forests their habitat. Biodiversity covers flora and fauna, and both are extremely bountiful in their diversity.

Forests, as a biome, are vital to Earth's health. While a rainforest is busy about many services to their local community, they will not easily be replaced by any human intervention as the best human ingenuity could not replace the work of a rainforest. Trees continually grow towards the sun and absorb their energy from the sun. At the same time, trees absorb carbon dioxide from the atmosphere, process it, and return oxygen to the air. The absorbed carbon is stored in the trees. It is stored carbon in fossilised forests that died millions of years ago that is now being drawn out of Earth as coal and oil and burnt back into the atmosphere. Rainforests are ultimate recyclers as nothing is wasted in forest life. As well as providing a habitat for a myriad of wildlife, rainforests store large amounts of water and are a very important part of the hydrological cycle, the water cycle that supports the world, not just the area where the forests grow as the water cycle moves water around the Earth. It can be argued that rainforests are a treasury of biodiversity, initiators and dispensers of fresh water, vacuum cleaners for air, and global heating sensors and regulators. Humans cannot replace the work of a rainforest.

Rainforests and forests in general around the world are under threat from logging for timber or clearing land for agriculture or the grazing of animals. Agriculture is responsible for about 80 per cent of tropical deforestation. Deforestation involves clear felling – that is, every tree is chopped down, the timber removed, and the remainder of the trees are usually burnt. Therefore, when it rains, topsoil can be washed away, polluting rivers and creating grasslands which, without reliable rain, can turn to deserts, as has happened in parts of Africa. In Australia, environmentalists are constantly defending the old growth forests and other forests from logging. It is thought that whatever we will do when forests are gone, we should do now and leave the forests to do their thing in maintaining the health of air and water for today and future generations. What is important is that we educate ourselves about the forests of the world and their unique gift to life. For example, if countries leave their forests alone, they will be healthier and wealthier for making that decision now for the future. Short-term commercial logging may benefit this generation, but future generations will be the poorer for inconsiderate choices today. If countries will make a decision now for the future security of their forests, they will have clean air and a healthy water cycle, and tourists will be able to flock from around the world to see the wonder of an old growth forest. What is important for us as people wanting to understand how forests impact on the health, balance, and harmony of Earth is to get to know what is happening to and in the forests where we live. Are they safe from human interference?

What, in a nutshell, is happening to forest biomes?

- It has taken hundreds and thousands of years for rainforests of the world to become established as macro ecosystems.
- Humans have an insatiable need for timber, so all rainforests are at risk.
- Over half of rainforests and forests across the world have disappeared over the last fifty years. This has largely occurred for timber and agriculture or been lost in wildfires.
- Every minute a hectare of tropical rainforest is destroyed somewhere in the world. How can we get our minds around that statistic? This is a massive, gigantic, and unimaginable loss to the health of Earth, us earthlings, and vital biodiversity.
- Because of the Technological Revolution, trees can be cut down at an extraordinary rate, and there's no stopping reckless people who are more about the dollar than the continuation of life on Earth. Sadly, native people, many who have given their lives to protect their forests, are fighting a losing battle against governments and guns. An interesting and encouraging story is to follow women in India who are defending their forests with their lives (see 'Standing Up for Trees: Women's Role in the Chipko Movement').
- Global warming generating a change in the climate may prohibit the renewed growth of forests from ground cover to canopy. The microorganisms in the soil needed to support trees may not respond to new growth as conditions will have changed for the soil as well. There

is so much more to forests than just the trees. Forests are a massive, interconnected network of life from soil to canopy, called symbiotic relationships in nature.

- Given that tropical rainforests are considered to be the 'lungs of Earth', we know that this is a vital role they provide for the continued health of the air we breathe. This service is being lost to the rest of the biosphere.
- Rainforests are burned every day either by design or by wildfires. As they burn, they release their captured carbon into the atmosphere. Wherever we are on the globe, we are breathing this polluted air sooner or later. Forest wildfires, usually caused by lightning, have been so intense that the forest floor is scorched, and native trees may not recover. Firefighters in Australia and California have experienced unprecedented temperatures as they fight forest fires. Is this the new normal for wildfires?
- Amazon Rainforest deforestation and dieback because of drought or acidic rain is one of the 'tipping points' that can greatly change how Earth functions as a living system. This is a 'tipping point' to watch.
- No wonder forest scientists are in despair – because they know the domino effect of the consequences of destroying tropical rainforests. This is truly a case of not only short-term gain for some thoughtless people but long-term loss to the human species as well as the extinction of biodiversity that live in the forests. There will be no coming back for many of them. All of us are the poorer whenever a creature that has shared our evolutionary journey is lost, even if it is a tree.

It is over fifty years since the 'Save the Rainforests' campaigns have been underway. We can ask ourselves, 'Do I care what happens to the rainforests? If I don't care, why don't I care?' It is a problem of education in joining the dots that clearly show we are all, including trees, interlinked. In downtown Melbourne,

I depend on the Amazon, the boreal, and the Borneo forests to do their thing in purifying air and keeping the worldwide water cycle in harmony and balance. Sitting here, writing this account, I have to say to myself, 'I need to care about the rainforests, and I need to beat my breast and ask forgiveness from all the creatures that have been lost to extinction on my watch through logging.'

I once went on a school excursion to a clear-felled area of forest in the Otway Ranges in Victoria about twenty years ago. I can say that I felt a deep sadness in my heart that, to this day, has not dissipated. The remains of a murdered forest – as in a few small branches and leaves, remnants of their rich diversity – lay before me. What I was looking at butted up to a tall vibrant forest family of trees and their biodiversity standing in line to be engulfed in the next weeks' slaughter. It was a very sad sight, and I felt a deep melancholy in my spirit. Twenty years later, I still feel that pain, that sense of loss, like my kin had died.

In destroying forests, we are not just 'biting the hand that feeds us'; we are also literally devouring the hand that feeds us. If I were a rainforest, what would I be saying to those who cannot see the potentially fatal outcomes of their actions? We have a long way to go to get the deep interlinked/interdependent ecological wisdom so imperative today. In general, we just don't get it. The question is are we eco-illiterate or simply imbued with a sense of superiority that we think we can destroy the natural world and it will be very magnanimous in its response?

Taiga Biome

Taiga is a Russian word for 'forests'. The taiga has traditionally been covered in trees that grow in ice and snow. The diversity and complexity of Earth's biomes is amazing and difficult to imagine or comprehend without experience of them. The taiga biome is in the Northern Hemisphere – a long way from my deck! It is that area below the tundra and above grasslands in some countries and temperate forests in others. Around the world, such places as Scandinavia, Canada, Siberia, and Alaska enjoy the taiga as part of their landscape. It is thought to be the largest land biome on Earth as it stretches across Eurasia, Canada, Russia, and North America.

During the Ice Age, the taiga was covered in ice and glaciers, but today much of the taiga is forested. Thirty per cent of world forests are located in the taiga regions. The average temperature is below freezing for at least six months of the year. The adaptation of the biochemistry of the trees has evolved to prevent them from freezing. Much of the soil is permeated with permafrost and constantly frozen. The taiga is known for its short summers and very long cold winters, and it experiences very low rainfall throughout the year, with precipitation usually in the summer months. Otherwise, snow covers the ground for a large amount of the year.

The taiga supports a wide range of biodiversity. As in other ecosystems, flora and fauna have adapted to cold conditions over centuries. Adaptation measures include insulating fur or feathers and, alternatively, migration or hibernation. Permanent residents such as hares seem to generally reside in nooks and crannies or holes in trees like squirrels. Larger predators – for example bears, Siberian tigers, and lynx – can be found alongside their prey, usually moose and deer. The taiga has very few native plants, just a few flowers or shrubs. Mosses, lichens, and mushrooms also find a habitat on the forest floors. Insects abound in the summertime, and birds flock to the taiga to gorge on them and breed.

The taiga has enjoyed millions of years of peace and tranquillity, but like other biomes, it is now under threat. Hunters and gatherers have been present for many thousands of years and lived off the land. Fauna provided food and clothing as well as pelts, skins, or hides for shelter and bedding. Human presence was part of the balance and harmony of nature because there are few signs of their habitation on the taiga. Today much larger threats to the taiga biome are evident.

There are two major threats. First, warming temperatures pose a real threat to the health of the taiga biome as the ice-clad taiga is extremely responsive to the rise in global temperature. As snow and ice melts, this can generate flooding in the taiga biome. As permafrost melts, it releases captured carbon and methane into the atmosphere that, in turn, increases Earth's temperature and consequently results in a feedback loop. Second, clear felling, logging for timber, destroys native forests and the habitats of creatures great and small. Huge swathes of trees are clear-felled for their versatile softwoods that are sent around the world. The deal is that as the trees are felled, replacement trees are to be planted, but it is unlikely that the native trees could return as environmental conditions are changing all the time. The conditions that native forests originally flourished in are not guaranteed into the future. The boreal forests absorb huge amounts of carbon as one of their great contributions to a healthy Earth. What is important is that we understand the contribution the taiga biome makes to the totality of a healthy Earth ecosystem. What happens in the taiga biome influences the tropics. It is important to remember that every ecosystem is linked, one to another, and they influence the health of one another.

What, in a nutshell, is happening in the taiga biome?

- Temperatures in the taiga are rising due to global warming.
- A warming climate contributes towards the thawing of the permafrost in the taiga. As temperatures keep rising, the melt will continue and worsen. The ground will become soggier and squishier.

- Boreal forests are one of the ecosystems most affected by climate change.
- The boreal forest is on the taiga biome, so like every forest, some people can't see the trees for the timber. This is a classic case of environmental health being at odds with economic prosperity. There will be no winners in this debate.
- Deforestation is a huge environmental issue for the taiga. The trees of the boreal forest are virtually there for the taking. They are a resource to be plundered at will. Perhaps when the forests in the taiga are gone, there will be no coming back because the cold climate that allowed them to grow in the first place may be changed with global warming.
- The forests are also susceptible to acidic rain because of pollution in the atmosphere, which, of course, affects the trees' health and can cause them to die.
- The clear felling of trees leaves the soil very vulnerable to not only erosion caused by rainfall but also the thawing of the permafrost. The roots of trees keep soil together.
- Other threats to the taiga are mining, drilling for oil, poaching native biodiversity, and human intrusion. All, of course, have only been made possible by modern technology that has allowed access to the taiga.
- Any significant change to the boreal forests is considered by climate scientists to be one of the 'tipping points' that can be irreversible and change the climate.
- Forests are a natural carbon sink, but with the impact of global warming and deforestation, they can change to being a significant source of greenhouse gas emissions.
- There have always been wildfires from lightning strikes, but research modelling is that there will be more wildfires expected because of the warming climate in the taiga.

The taiga is a magical and mysterious place. It is an absolutely beautiful wilderness biome. It is home to its own biodiversity, such as the Siberian tiger, which is on the endangered list because of the loss of habitats and poaching. This is the 'land of drunken trees'. When the permafrost thaws, the trees sink and begin to lean over. The taiga is one of the last frontiers for exploitation; gone are the days when everything and everyone lived in harmony on the taiga biome. In the natural world, every action has a reaction, so how prepared are we for Mother Nature's reactions to upsetting the balance and harmony of the taiga?

In summary, the geographical and climatic biomes cover the entire landscape. Each biome contributes to the health of the whole Earth. Just because we do not understand the importance of the services they provide does not mean we can dismiss what we do know. Sadly, every biome is threatened by the human species. There is no place on Earth that is safe from human interference or exploitation. Human progress depends on a willingness to exploit every possible resource/gift of Earth. Sadly, this is done without too much conscience about how a depleted Earth will impact on future generations of the human species.

One thing for sure is that evidence is already available regarding the impact that the human species is having on the biomes of the world. Global warming influencing climate change, along with the degradation and exploitation of natural resources in all biomes, is setting up a devastating scenario that will ensure that millions of creatures are endangered or, at worst, threatened with extinction. We must understand that we are the poorer as humans if any creature goes the way of extinction after journeying through the evolutionary process with us. If this message sounds like a recording to you, then be assured that it is an extremely repetitious statement in the hope that we will grasp the seriousness of our dilemma if we continue to push Mother Earth to the point where we will experience retaliation from nature because Earth is absolutely responsive to our behaviour. The biomes of the world are big-picture stuff that we do not reflect on enough, but we do need to be aware of what is happening and how fast it is happening so that we can anticipate feedback loops tending to tipping points that will be catastrophic.

CHAPTER FIVE

What, in a Nutshell, Is Happening to Ecosystems?

What we know is that within each biome, there are bountiful ecosystems. There are a number of life-giving ecosystems to reflect on. When people talk about the environment, everything is usually lumped together so that we do not get a clear picture of what is really happening to the interlinked ecosystems on which all life depends. The plight of ecosystems does not make easy writing, and therefore, it will not make easy reading.

We know in our hearts, if not in our minds, that something serious is going on with the health of Earth, even if we cannot see or feel it. We hear snippets of information about the degradation of ecosystems in the media, but we seem to be incapable of piecing the information together so that it makes enough sense to call us to action. In my thinking, environmental degradation is as serious a threat to humanity as global warning and the changing climate. It is a similar challenge in trying to come to grips with the complexity of consequences resulting from anthropocentric global warming. We know Earth is heating up incrementally every year, and we know why, but we do not, in general, understand how that affects us in so many ways – with food security, for example.

A brief study of each ecosystem allows us to at least pose the important questions and hopefully get a holistic view of the role that ecosystems have in maintaining a balanced and harmonious Earth so that we feel empowered to make the necessary changes to how we live, act, and vote.

Ecological conversion is founded on understanding how Earth functions and thoughtfully examining the present state of Earth regarding its health. To do this, we can look more closely at the ecosystems supporting life – that is, air, water, soil, and ocean – and step outside our own backyards and familiar environs to see what is going on with the health of Earth's ecosystems in the unfamiliar. It is also valuable to get a picture of species that have shared our evolutionary journey to see how they are doing in this time of rapid environmental change.

In general, I think it can be said that we are very anthropocentric in our thinking. We are pretty much just concerned about what affects us on a day-to-day basis. What I mean is that all threatened species have their stories to tell, but they have no voice but ours. You may have heard Dr Jane Goodall talking about her life with the chimpanzee families, and she is now pleading with the people of the world to protect them, especially as they are our closest relatives, something like 98 per cent like us. To become Earth literate and undergo an ecological conversion, it is imperative for us to understand how Planet Earth, our home planet, functions.

Not only is the Earth the provider of all the resources/gifts we need to live, but also, we have to be attentive to all the 'wisdoms' imbued in our planetary home. If we are going to join the dots to understand what is going on with our home planet, then we first need to appreciate our relationship with Earth. Mother Earth is our primary educator; Mother Nature is our best teacher, and it is so important to work with her rather than ignore her. It is essential to sit attentively at her feet and pay attention to her voice within each of us. The Earth and its health can best be assessed by looking at individual ecosystems while remembering that all are interlinked, interconnected, and interdependent. What goes on in one ecosystem influences the health of every other ecosystem – that is, all are intertwined/interlinked with one another.

Ecosystems are well-defined biological societies made up of biotic and abiotic creatures. No ecosystem is an island or acts independently. Biotic creatures are extremely diverse in their expression – that is, they include plants, animals, insects, and every living creature. The abiotic make up Earth's elements, such as air, sunlight, soil, and water. Soil and water, for example, are full of microscopic life. So what can we learn from Mother Nature as we explore various and multiple ecosystems? The following reflections offer a brief introduction to thirteen of them; all, of course, are interconnected, and each contributes to the health of Earth in its own activity.

Aquatic Ecology

Whole books, perhaps whole libraries, have been written about aquatic ecology. There is no shortage of information about water in all its forms. Water links all five biomes; it is an infinite gift of Earth. The water on Earth gathered during Earth's formation many, many millions of years ago. How water came to Earth is still something of an enigma, one that you can explore. Water is what made life on Earth possible as life began in the ocean. That may explain why our sweat and tears are salty! Without water, Earth would be a barren planet like Mars, and most creatures would not have evolved and survived their evolutionary journey. Water is a mystery and miracle, but we do tend to take water for granted. Earth is a 'water world' planet as water covers about three quarters of Earth. It is held in the ocean, rivers, lakes, wetlands, glaciers, ice, artesian aquifers, clouds, and every creature. Every splash of water is full of life. Under a microscope, you can see a myriad of life in a single drop of water.

Water ecosystems are divided naturally into fresh water (that is, rivers and lakes) and marine water (that is, salty, like the ocean, ocean estuaries, and coral reefs). Different species of flora and fauna are adapted to either fresh water with extremely low salt density or marine water with a very high concentration of salt. However, some creatures, such as some birds and crocodiles, are adapted to both types of water.

Freshwater ecosystems such as lakes, rivers, creeks, ponds, wetlands, and manmade dams provide a habitat for millions of aquatic creatures. Creeks and rivers can be found everywhere. Some flow constantly because they are fed from rainfall on mountains or melting snow and glaciers. Rivers follow natural terrains and eventually flow to the sea. Wetlands known as marshes or swamps are usually stagnant or still water and are fed by flooding from ocean surges or a deluge of rain. Coral reefs are found as barriers out from the shoreline along continents or islands. Estuaries are usually located at the mouths of rivers, where they meet the sea. Fresh water meeting salt water provides a very interesting ecosystem.

Aquatic ecology is not without its environmental issues because of human activity; for example, wetlands are being drained for housing or agriculture, and many rivers no longer reach the sea to cleanse themselves because of an excessive demand for irrigation of crops and domestic consumption. The changing climate and pollution are impacting on the health of aquatic ecosystems, resulting in a subsequent loss of services that they perform to ensure a healthy Earth. The aquatic biome links all ecosystems because like air, water moves around the planet. What we can do to help is learn about aquatic ecology, its traditional way of life, its systems, and its flora and fauna as well as appreciate its essential contribution to life on Earth. We can be a voice for rivers, lakes, ocean, wetlands, and coral reefs by getting informed about our local water ecosystems and doing what we can to clean and protect them.

What, in a nutshell, is happening to the aquatic ecosystems?

- Every aquatic ecosystem has been degraded by human activity.
- The fact that so many rivers no longer reach the sea is an extremely concerning environmental issue because that means the rivers are not able to flush themselves. As a result, their water can become stagnant and therefore unhealthy for the creatures that live within their watery habitats.
- Human demand on many aquatic systems is so great that they are not able to function in their traditional way.
- The rivers that are fed by glaciers or melting snow are affected by global warming, and into the future, there may not be enough melt to service the rivers in their normal way.
- As global warming heats the rivers and waterways, the creatures that live in them will not survive as all creatures have adapted to a specific temperature.
- Global warming will also accelerate the evaporation of water ecology.
- The pollution of rivers through agriculture runoff, human negligence, or flooding is a major source of concern, especially as the rivers flow to the sea and deposit pollution, which is a death trap for many creatures.

Environmental scientists are justly alarmed about the threats to a healthy aquatic system. Much of the degradation of water systems has occurred in the last thirty years as more and more demands have been made on them. As the human population increases, demand will naturally increase, and so what is already degraded will be threatened to the point where aquatic systems can no longer provide the services of the past. The writing is on the wall as many creeks and tributaries no longer have flowing water, and many in Australia are bone dry all year round now. Each of the aquatic ecologies will be looked at more closely in this chapter.

Ocean Ecology

For millions of years, rain fell on Earth; it was water that enabled life to come into being. Over time, water subsided to the lowest land, forming the ocean. There are five great seas on Planet Earth, and they all join together to make one amazing and awesome ocean. The ocean covers about two-thirds of Earth's surface, and so from space, astronauts are able to see the great Pacific Ocean, which virtually covers one side of Earth. Some astronauts thought Earth should have been called Planet Water from the life-changing view they had of Earth. As the astronauts watched Earth rise from behind the moon, they said Earth hung in space like a beautiful jewel against a black backdrop. The photos that the astronauts took as they came around from behind the moon have changed forever the way we perceive our planetary home.

The ocean is the largest ecosystem on Earth. It is the largest biome on Earth as it covers 70 to 75 per cent of Earth. It was home to many billions of wild ocean creatures. It has microscopic creatures and the largest wild creatures on Earth that are alive today. The ocean's biodiversity is rich and so numerous that the creatures cannot be counted. What we do know is that the ocean provides food not only for human consumption but also for the many billions of creatures that live in the seas. Underwater photography has made it possible for our generation to see below the waves and wonder at its beauty and diversity.

Ocean water is salty. For millions of years, as water was drained from the land, it carried with it salt and minerals dissolved from the Earth's crust. The composition of salt and minerals in the ocean has not changed over many millions of years until recently. Human activity is releasing tonnes of carbon dioxide and other chemicals into the air. The ocean is sometimes called the 'blue lungs' of Earth. When it rains, raindrops join with chemicals in the air and deposit them into the ocean as toxic rain, thus making the ocean more acidic. This process is referred to as the acidification of the ocean. The polluted rain, therefore, is changing the chemistry of the ocean, which is extremely harmful to many fish, seaweeds, and especially coral reefs. We eat fish, so that is not good for the human species either.

As well as acidic rain from polluted air, humans have also used the ocean as a sewer. Many towns and villages have sent their sewage out into the ocean on the tides. Chemical waste from factories has also found its way to the sea as a convenient means of disposing of it. Storm water from cities is also an issue because water in drains collects rubbish – for example, cigarette butts, plastic items, and all manner of human waste – and rubbish goes out to sea along with the storm water. Oil spills are another environmental issue. Dead zones have appeared after massive oil spills, a serious hazard to ocean wildlife. Also, ocean-going ships have thoughtlessly dumped their rubbish in the sea for decades. They thought 'over the side' was away, but there is no such place as 'away' on Earth. All their rubbish is now breaking up on the sea floor by wave action or is being ingested by unsuspecting creatures that don't know human rubbish is deadly. No one thought about the long-term impact on sea life as we were doing our dumping, but it has come back to haunt us now. We cannot sieve the ocean and retrieve our rubbish.

Since the advent of plastic, about sixty years ago, enough plastic has found its way to the ocean to form huge plastic islands of rubbish captured in the centre of gyres. A gyre is that space in the ocean that does not have strong currents to move it around. The Great Pacific Garbage Patch can be seen from outer space as this pile of plastic garbage that is so huge, it appears as an island. As the plastic breaks down into smaller pieces, it forms what scientists call plastic soup, but a further breakdown of plastics results in sinister and deadly microplastics, which we are all unintentionally consuming. Microplastics have even been found in unborn foetuses. Many fish and birds fall prey to plastic as they cannot digest it, but they keep on eating it, thinking it is food, and sadly, birds feed bits of plastic to their young chicks. The remnants of decayed chicks have been discovered, and a coroner's report would say 'death by plastic'. Recent scientific research has shown that even microscopic creatures such as plankton have ingested plastic. People like to fish, we love ocean sport, and we love beaches, but now we need to be rethinking how we can contribute to the health of the ocean and all creatures that call the ocean home.

The ocean has a significant role in regulating the climate; however, climate scientists are concerned that global warming is heating the ocean as never before in human history. When water is heated, it expands, so ocean levels are rising and therefore threatening islander peoples' homes and coastal cities. Climate change is the biggest single threat to ocean health, but it's not the only one. Ocean warming influences coral reefs that have grown in a stable water temperature for millennia and are extremely sensitive to any change. Ocean warming also impacts on the icecaps of both the North and the South Poles and subsequently is causing ice to melt from below the waterline, adding more fresh water to the ocean.

What, in a nutshell, is happening to ocean ecology?

- The health of the ocean is imperative for all life because all ecosystems are interlinked.
- Ocean biodiversity is threatened by global warming. Wildlife in the ocean has adapted to temperatures that have allowed them to thrive over millennia. Ocean wildlife do not handle change very well if at all.
- The world's oceans are heating up as much as 40 per cent faster than scientists previously thought was possible. Since the Industrial Revolution, the burning of coal and exhaust from petrol cars, diesel trucks, airplanes, and white goods has been absorbed largely by the ocean, possibly as much as 85 per cent of excess heat in the atmosphere.
- The global warming of the ocean plus the melting icecaps are raising ocean levels, impacting on housing and infrastructure. Some islander people are losing their homes to the sea. Icelandic people are having to move their villages further up the ice as they are being threatened by the rising oceans because of ice melt. On the east coast of Australia (2020), houses built with ocean water frontage are being eroded and perilously in danger of collapsing into the sea, and some will need to be demolished because of instability. El Niño was blamed for the event, but the houses were not built yesterday, and many El Niños have come and gone without mishap, so one is left wondering. There are possibly many examples of the impact of global warming influencing sea levels around the world, but like a lot of environmental issues, they can be mitigated by adding rocks, pumps, sea barriers, and walls today. Still, will all be well into the future?
- Rising temperature in the ocean is impacting on ice at the poles – that is, the Arctic and Antarctica. Ice, which is fresh

water falling into the sea because of ocean warming, is changing the chemical composition of the ocean.
- The rising ocean temperature affects coral reefs, the spawning habits of fish, and kelp forests and promotes algae blooms, amongst other consequences.
- Pollution in the ocean as a result of chemical runoff is creating havoc for the wildlife of the sea.
- The Great Pacific Garbage Patch and other gyres that have collected mountains of human debris are visible from space. The plastic is breaking down, forming plastic and microplastic soup, which becomes fatal food for wildlife in the ocean. Hundreds of whale and shark carcasses have washed up onto the shore, and when they are examined to ascertain their cause of death, many kilograms of plastic are found in their stomachs. We are left to wonder how many more have died and sunk to the bottom of the ocean or been eaten by other creatures of the sea.
- Acidic rain is changing the chemistry of the ocean.
- Oceans are now subject to deoxygenation as global warming and pollution inhibit plant growth such as kelp. Algae is replacing traditional ocean flora.
- Overfishing is changing the diversity of wildlife in the ocean. As each large fish is removed from the ocean, the biodiversity will change, especially as we consume each new level of fish stocks. We can expect more shark attacks as they need to come closer to the shore to feed. Humans have eaten their traditional food. Every action has a reaction.
- The ocean was the largest habitat for biodiversity on Earth. The richness of life in the ocean and the diversity of wildlife impossible to count.
- The depletion of 70+ per cent of large ocean fish is due to overfishing and the immeasurable loss of many other fish that are lost through 'by catch'. Humans are devouring

ocean wildlife to the point where some people are down to eating jellyfish!
- How the ocean functions as an Earth system is cause for consideration as it is one of the potential 'tipping points' that could be irreversible – that is, irreparable – once it has been influenced by global warming, causing rising sea levels and also freshwater intrusion from melting ice.
- Ghost nets abandoned by fishermen are the principal cause of loss of life in the ocean after ravaging by the human species. Ghost nets continue to trap and drown many wildlife creatures. It is hard to digest that dolphins or turtles actually drown because of human fishing pollution. A walk along the seashore will confirm this disaster as lines and nets readily wash up on the beaches.
- The loss of biodiversity through by-catch and killing sharks just for their fins is a crime against creation.

With all this going on, is it any wonder climate scientists are nervous about the continued health of the largest ecosystem on Earth? They are also deeply concerned about the possible changing role of the great ocean conveyor belt and how it will adjust to changing environmental conditions like global warming, influencing freshwater intrusion from melting icecaps and glaciers into the ocean. In nature, every action has a reaction; remember, everything is interlinked. Ocean currents are influential in determining climate and weather conditions around the globe. They are constantly moving and circulating warm water, which is less dense around the globe, which then influences weather; therefore, changes in ocean water temperature can be a potential 'tipping point' for environmental catastrophe. No wonder climate scientists are scrambling to understand all the impacts of global warming on ocean currents! The ocean is extremely complex in its activities and the services it provides to the ecosphere. Humans have never considered that their behaviour could influence the health and activity of the ocean, but they have.

As land dwellers we don't think about the need for a healthy, balanced, and harmonious ocean, but we are upsetting that balance that took many millions of years to evolve. We stand on our beautiful beaches, but we cannot see any of this happening; however, it is happening, and we are the culprits. We are devouring ocean wildlife and destroying ocean biodiversity at an unsustainable rate. Island nations confirm that the fish they have lived on for generations are no longer there for them, so they are now eating the next level of fish available to them, but when they are gone, where do they turn to for their food since fish is their staple diet? The thing is do we care enough to take a stand for the health of the ocean and its biodiversity? Can we join the dots and see that we need to be proactive and support the many calls of climate scientists to change our thinking about the ocean and be a voice for its preservation?

This is a very hard read. As humans, we do not want to think about the health of the ecosystems that support our lives. We do not like to view ourselves as plunderers of Earth's riches; we just want to sit on the beach and think everything is good with the ocean's health. We can feel powerless to do anything about it as it all seems to be out of our control; however, I will write later about the 'power of one' and the 'power of the people' to make a difference, so hang in there. Some say writing things down is good therapy, so you might like to take on the persona of a sea turtle since they live a long time. What has been your turtle experience of ocean life? How has life in your ocean wilderness changed?

We are, in general, not smart enough to understand that the health of the ocean's harmony and balance is absolutely paramount to our existence. We are pretty much incapable of comprehending all that the ocean, with its biodiversity, does to keep the whole Earth stabilised and synchronised. We can think that our response is about the fish that we eat or concern for all the wildlife of the sea, but it is so much more as the ocean influences our climate. If we lose the services of the ocean, then our own lives are in deadly peril. Without education, we are going to continue to allow ourselves to degrade or even destroy this absolutely vital life-sustaining ecosystem. I recently heard a marine scientist say that if the ocean dies, we will die with it. I thought that sounded a bit harsh, so I did some research and sure enough there is a large following of that theory. The plain fact is that the services that the ocean provides are vital for life on Earth.

The International Program on the State of the Ocean (IPSO) studies the ocean to educate us on its vital role in maintaining the life support systems that have allowed life to exist. Although humans have named different seas and oceans, they are linked in one awesome global system. If forests are called the 'lungs' of Earth, the wetlands the 'kidneys', and the soil the 'skin', then the ocean is Earth's 'blood circulation' system. The ocean is not static. It has a great conveyor belt that transports water from the depths to the surface and vice versa as well as moves warm and cool water all around the globe. When we fully understand how the ocean functions and what is happening to the health of the ocean as an environmental issue, then we are in the best position to decide to be a voice for the ocean. It is quite clear that the health of the ocean requires many voices, so we need to start yelling!

Air Ecology

All living, breathing creatures need air, but again, we tend not to think of other creatures that also need fresh, clean air. Earth's air was originally dominated by carbon dioxide, which made life on Earth impossible except for cyanobacteria or blue-green algae. These amazing creatures took in carbon dioxide and released oxygen. Over millions of years, slowly but surely, enough oxygen filled the air to make life possible. It is one of the most important building blocks of life.

Air is an infinite gift of Earth because it is continually recycled. It is an interesting phenomenon because we cannot see air, although we can feel it when the wind blows. We have to breathe air as we cannot live for more than a few minutes without it. We do not think much about the air we breathe; even in our sleep, we just go on breathing in and expelling air. We share the air with every breathing creature, and we are constantly breathing air that has been exhaled by someone else or some other creatures, like whales. Air binds us together in an extraordinary way. It binds us to the past because we are continually breathing air molecules that dinosaurs or Neanderthals breathed in and exhaled.

Air is made up of about 78 per cent nitrogen, 21 per cent oxygen, and less than 1 per cent carbon dioxide. It exists in a very delicate balance for life as it only extends out in the troposphere for about ten kilometres. People who climb Mount Everest have to take oxygen tanks with them sometimes. Commercial aeroplanes must be oxygenated because the air is so thin at the heights they fly at.

We used to think that breathable air extended out into the universe, that it was unlimited. Air is one of those amazing gifts of Earth that is continually recycled through nature's awesome air cleaners. Plants take in air, use the carbon dioxide, and release oxygen through photosynthesis; people and animals take in air, use the oxygen, and release carbon dioxide. What a great reciprocal partnership! The forests, ocean, and soil are also part of nature's air cleaning system. This is an ecosystem that humans cannot replicate, so it is vital that we understand the importance of clean, fresh air for our health and well-being.

There are many studies done on the impact of polluted air on the human body and countries that are breathing in harmful particles from smog conditions. Some countries are setting about cleaning the air they breathe because of an increase in lung disease and other poor air quality related ailments. Today, because of modern technology, we are actually changing the composition of air, which is a serious ecological concern. This means that the chemistry of the air is changing because toxic pollutants are putting huge pressure on natural air-cleaning processes. Ecosystems are struggling to recycle air as they have done for eons. Quite clearly, the pollutants that have been released into the air not only generate global warming and produce a health risk but also have upset the balance and harmony of the air in the atmosphere. Added to an increased volume of toxic pollutants in the air largely because of burning fossil fuels, humans are destroying air-recycling forests at an incredible rate. When forests are cut down, what is not suitable for use is burnt, thus releasing their captured carbon back into the atmosphere. Trees absorb carbon, and that is why wildfires are so bad – because as they burn, carbon is released, and it returns to the air from whence it came. No air is lost to Earth, and no air is added to Earth; air is constantly recycled. Hence, it is imperative that every person does what they can not only to limit toxic pollutants being released into the atmosphere but also to plant vegetation, trees, or veggie gardens to absorb and sequestrate carbon dioxide.

We do not think about air all that much, but we need it regardless. When we inhale air, our lungs absorb the oxygen to fuel our bodies. Oxygen is extracted and passed into the bloodstream and delivered to our tissues, organs, and cells. Each cell requires oxygen. When we exhale, we release gases, including carbon dioxide, which is the work of our respiratory system. We depend on fresh, clean air for a healthy body. We are air!

What, in a nutshell, is happening to air ecology?

- Air is made up of carbon, nitrous oxide, nitrogen, methane, and other greenhouse gases.
- Human activity is changing the chemistry of the air.
- The rise in the quantity of pollutants in the air is creating potentially catastrophic global warming, which has consequences for every ecosystem.
- Pre-industrial greenhouse gases in the air were 280 ppm because of Earth's natural functioning. Now air pollutants have reached 410 ppm, and the amount is climbing exponentially as 'business as usual' is not, in general, respectful of air quality.
- Burning coal, oil, and gas to service our lifestyles and provide energy for manufacturing has seriously exacerbated the amount of toxic chemicals in the atmosphere. Thankfully, countries are now weighing up the cost of destroying air quality that has intensified global warming of the planet.
- The Technological Revolution over the last five decades has exponentially increased pollutants in the atmosphere to the point of altering the harmony and balance of air quality and composition of gases.
- Smog, which lingers over cities, is made up of dangerous chemicals that can create dark fog. We who live in cities don't realise that we are breathing in heavily polluted air. It is not until you drive out into the country that you realise the air is so much clearer and cleaner and even smells different.
- Nine out of ten people breathe air that has high levels of pollutants, resulting in health issues such as asthma, premature deaths, the increase chance of stroke, heart disease, lung cancer, respiratory infections, birth defects, and reproductive issues.

- It is a saying that the 'devil is in the detail' well there are 'toxins in the dust' in our homes. We are just not aware of this fact, but they are identifiable and measurable!
- Natural air pollutants are from volcanic eruptions, wildfires, and allergens such as pollens and mould, but nature can take care of their contribution.
- Polluted air is a silent killer for man and beast.
- Human activity to produce energy, agriculture, and industrial incinerators are the principal causes of the rise in pollutants in the atmosphere.
- Air pollutants produce acid rain, which effects soil, vegetation, the food we eat, and ocean acidification.
- One issue we do not think about at all is the impact that air pollution is having on buildings and infrastructure, such as weakening, eroding, and potentially destroying them over time.

Now with all that information about human-induced chemical changes in the air and the impact it is having on our personal health and the health of the planet, how can we fail to act on what climate and environmental scientists have spent their lives studying and measuring? Everyone and every creature are interlinked by air that moves around the globe. Is it that we do not understand the reality of it, or is it really a matter of faith? Do we choose not to believe that we are capable of polluting the air we breathe, to the detriment of our own health, let alone holistic planetary life? We quite simply cannot go on polluting the air without it having detrimental consequences for the health and well-being of all life on Earth.

Soil Ecology

Just plain old dirt, aye? No. Dirt is the stuff of life; it is extremely precious. It is also called 'earth', 'ground', and 'soil'. Soil consists of organic and inorganic matter and varies in chemical and physical properties. Most importantly, it is alive. To my way of thinking, soil is the Cinderella of the three sisters – that is, air, water, and soil. If we don't think much about the air we breathe or the water we drink, then we are not likely to think at all about soil. Soil produces most of the food for most of the Earth's land-loving community of beings. From soil come the plants we eat, and soil produces the plants that animals eat, and we eat the animals, so soil is good. We eat soil! If we have a fish-free meal, then everything we eat comes from soil. Just as we are air and water, so are we soil. We are literally soil people.

Soil, in its many forms, covers most of the landmass. It has taken billions of years for volcanic rocks to break down through wind and rain to decompose into what we know as soil. Remember, in the beginning, the Earth was covered with very hot melted rock. Soil is that top layer on land; in fact, it is called topsoil because it covers the Earth's landmass like a skin and is not very deep. In fertile valleys, soil can be very deep, but in places like Australia, it can be a whole lot less. It certainly does not go to the centre of the planet. However, just as soil is being constantly blown away by the wind, so also is soil continually added to by the erosion of mountains and the decomposition of plant and animal matter to produce humus. Soil is alive; it lives. A spoonful of dirt could hold hundreds of microorganisms.

Soil provides a habitat for millions of microbes and bacteria as well as many millions of small creatures such as worms, grubs, and beetles. All these tiny organisms and micro-organisms work together to break down minerals plus plant and animal matter to keep renewing soil. This is why composting organic matter is a necessity for healthy soil.

As an ecosystem, a macro ecosystem, soil is under duress. Healthy soils have been, for quite some time now, subjected to being sprayed with herbicides, insecticides, and pesticides of all kinds as well as other chemicals – such as pre-emergent, emergent, and post-emergent chemicals – to make soils produce bigger, better, and faster crops. While plant life may reluctantly respond to the use of chemicals, organisms beneath the surface of soil are often poisoned, and soil is depleted of not only tiny organisms but also the nutrients they would normally leave in their wake. People are generally aware of the good work worms do in breaking down organic material for their gardens. Acidic rain – that is, rain that is polluted by excessive carbon dioxide and other chemicals – also influences the health of soils. Acidic rain makes it very difficult for soil to keep its biodiversity healthy. Soil also assists in keeping the planet healthy through its water-recycling process, another of nature's great recyclers. Humans cannot make soil; it is another gift of Earth.

Soil has lots of challenges to stay healthy. The exploitation of soil through stupidity or greed has resulted in soil being denuded of nutrients in probably every country. There are a number of causes for despoiling soil, such as land degradation resulting from the overgrazing of cattle or sheep, infestations of feral animals such as rabbits, soil erosion by wind and rain if the native ground cover has been removed, the lack of respect for soil suffering from droughts, noxious weeds taking over native grasses, salinization because of irresponsible irrigation, and worst of all, the deforestation of native trees. Australia has a reprehensible history of clearing the land of native trees and grasses. Some farmers are just beginning to realise that to sustain and restore good, fertile soil, especially through drought and global warming, native tree cover plus native grasses are vital. Today farmers and scientists are working closely to reclaim lost soil by replanting trees and native vegetation to stop the progress of salinity and soil erosion. We are all invited to be soil people because soil nourishes not only our bodies but also our minds and spiritual selves. Hands love soil. Kids have always played in dirt until now.

What, in a nutshell, is happening to soil ecology?

- Global warming is giving soil grief because higher temperatures dry out soil, causing drought conditions.
- Nearly all soil is contaminated because of acidic rain.
- Excessive amounts of irrigation for agriculture are causing more soil to become saline. Salinity is a growing problem as saline soil, in general, will only produce salt-resistant vegetation, which livestock, in general, do not eat.
- Trees are being cut down or traditional vegetation removed, so if crops fail, as they do, then the topsoil can be easily blown away by the wind, resulting in less soil or more arid dirt. Dust storms are a menacing phenomenon to behold.
- Soil erosion through wind and rain, especially storms, washes topsoil away, usually into creeks or rivers if the soil is not blanketed with native grass.
- Poor farming practices in the past caused excessive amounts of topsoil to be lost from farms. There was an old saying that 'farmers went to the city once a year to collect their farms' – that is, soil lost to dust storms; such was the loss of topsoil in those days because of the multiple ploughing of land for crops. This practice has largely been outlawed today in favour of direct seeding straight into the soil without disturbing native grasses.
- Excessive amounts of synthetic fertilisers have been poured onto the soil, destroying nutrients in the soil and adding salts or heavy metals to the soil. This means the loss of a myriad of microorganisms and biodiversity in the soil.
- Urbanisation is another disaster for good, fertile soil. Houses are frequently built on soil that is most suited for growing food. Of course, asphalt, concrete, plastic lawns, and all kinds of soil-covering materials have effectively killed the soil underneath them.

Soil, air, and water are the stuff of life. Food security for the world's population of people and all creatures depends on healthy soils. Sadly, we have run roughshod over soil in our arrogance and, to some extent, our ignorance of what soil ecosystems need to be healthy and productive; we have in general, not shown soil the respect it deserves. The human species has tried to make soil do their bidding without cognisance of what soil itself needs to be healthy. However, the evidence is clear; we are quite capable of destroying the soil that feeds us. Can we join the dots that explain why soil is in trouble and do our bit to nurture and preserve soil for its own sake? Thankfully, there is a regenerative farming movement that is gathering momentum in the farming community to rethink the importance of maintaining healthy soils and native grasses for the future of their various enterprises.

Antarctic Ecology

Very few people would give the Antarctic a single thought. For most of us, it is a 'nowhere' place where penguins live and of little importance in our 'business as usual' days. However, the fact is that it is imperative to us, and we need to be attentive to what is going on down under. The Antarctic was one of the last frontiers for humans to discover its wonders. It is largely a huge ice continent situated at the South Pole in the Southern Hemisphere; it rises over two thousand metres above sea level, and it holds about 90 per cent of Earth's ice. Like the Arctic biome, it is farthest from the sun and therefore one of the coldest places on Earth. The mass of ice is almost twice the size of Australia and a kilometre and a half thick. Antarctica also holds about 70 per cent of Earth's fresh water. That is important to appreciate given global heating. Antarctica is huge; it is about fourteen million square kilometres and surrounded by the Southern Ocean. It is also the world's tallest continent. Every so often, a huge chunk of ice will break off from an Antarctic ice shelf. The size of these icebergs is hard for us to get our minds around. They are a fascinating study in themselves. Some calving is ice just doing its thing as usual, but climate scientists are always assessing the impact of global heating on the ice sheets, especially now as the surrounding ocean is warming.

No one lives permanently on Antarctica, although many scientists stay for extended periods because they want to understand and record life there. It also has a good number of visiting tourists who just want to experience it and observe its beauty. There are scientists who research global warming, glaciers, penguins, whales, seals, fish, and krill, a most abundant and important food source for the wildlife of the sea. They also study minute microscopic life in extremely sparse amounts of soil. The conditions for the composition of soil are not present because of extreme cold. On a more fascinating note, Antarctica is the meteorite capital of the world because dark objects, such as meteorites, are easily spotted on ice. Scientists who study the changing climate drill into the ice for core samples to examine. The long cylinder-type ice cores reveal much of Antarctica's past and provide evidence of human activity around the planet, such as carbon dioxide emissions that present in the ice cores as dust or air bubbles. Earth is one whole ecosystem, and what happens to air and the ocean is recorded in ice and water around the ice continent, so scientists carefully watch for any changes.

There are some very iconic creatures that call Antarctica home. Penguins are in continual residence. Penguins can endure incredible cold winters as they have learned to huddle together to protect one another so they can withstand ferocious winter winds. Krill abound and provide food for whales and many other wild Antarctic creatures. Scientists are also researching commercial fishing, and they give regular updates on the physical conditions of penguins and their reproduction rates as possible indicators of humans overfishing the food supply of penguins, seals, and inhabitants of Antarctica.

There are some tiny ice-free areas on Antarctica – that is, about 0.44 per cent to 2 per cent – prime real estate for humans and penguins, and believe it or not, some plants grow there. The plants that grow there are ground cover, such as lichens and mosses that flourish in niches that provide protection from wind and cold. Antarctica is rather unwelcoming for humans, but those who do visit as scientists or tourists expound on its wonders and remain captivated by its mysteries. It is a continent that inspires imagination. Most of us will only dream about going there.

What, in a nutshell, is happening to Antarctic ecology?

- Antarctica is an example of 'the big meltdown', both metaphorically and literally.
- There is not really a nutshell explanation for what is going on in Antarctica. It is such a massive continent, and Antarctica is experiencing a number of different issues, particularly on its peninsulas and western or eastern sides.
- Global warming is affecting Antarctica more quickly and decisively than on land. It is believed by scientists who have studied Antarctica for decades that the impact of global warming of three degrees Celsius is even worse than what is happening in the Arctic Circle.
- In recent times, scientists have experienced rain for the first time during their research on Antarctica. That is different.
- On the western side, which seems to be more influenced by global warming, six hundred glaciers are in retreat out of about seven hundred glaciers. That is a fairly convincing statistic, and there is no need for faith. The glaciers are visible and measurable; they can be photographed on land and from satellites.

- The western peninsula is the place to research the impacts of a changing climate. Glaciers are floating shelves of ice. A number have already thawed and crumbled into the sea, thus threatening a rapid rise in sea levels. Massive ice shelves have already broken off and are floating as icebergs in Antarctic waters. In August 2019, an ice shelf over six hundred square miles in area and about seven hundred feet thick broke off. Ice shelves have always broken off from Antarctica, but scientists are concerned about significant changes in their behaviour.
- There are a number of domino effects with Antarctica's biodiversity. First, Adélie penguins like to build their nests on rocks or soil. However, now there is snow, so their eggs are apparently drowning because of melting snow; it is calculated that breeding pairs have dropped by about 70 per cent. However, other penguins are moving into their traditional territory because of the climate being more conducive to them, the invasive species. Second, humpback whales are increasing in numbers significantly because the sea ice is slower in growing, which gives the whales an extra couple of weeks to take their fill of krill, to the detriment of food for penguins. In this environmental change, whales are winners, and penguins are losers. Third, the Antarctic silverfish, the sardine-like staple food for Adélie penguins, has almost disappeared; scientists think global warming has altered the water temperature, so the penguins are reduced to eating krill, even though they have far less nutrient value than small fish, and subsequently, their chicks are struggling to stay alive.
- Then add in the human species as some countries send fishing trawlers to catch the krill using nets that rake in many tons for omega-3 pills and other products.
- There is a huge amount going on in Antarctica and on a grand scale, so well worth keeping abreast of the science. In fact, what is happening on West Antarctica is considered

by climate scientists to be one of the 'tipping points' that can permanently change the ocean, especially ocean currents, and therefore the climate.

Antarctica is one of those environmental issues that scientists seem reluctant to talk about because what is happening is unprecedented and a huge threat to the health of the planet. Whatever happens to the Antarctic sea ice will eventually affect us. Climate scientists fear for the continuation of this virtually pristine continent as a wilderness icescape. The fear is that as the ice melts, land will be more accessible, and humans will become an invasive species who will drill and blow up stuff, threatening the end of a mysterious, majestic, and magnificent wilderness island. For those who have been to Antarctica, they say that it is the most beautiful jewel in the crown of Earth. To see Antarctica is to love Antarctica, and no doubt scientists who have fallen in love with Antarctica want it to remain in its own awe-inspiring solitude.

Arctic Ecology

At the opposite end of the globe is the Arctic at the North Pole. Unlike Antarctica, there is no land at the North Pole. The Arctic is made up of a thin sea ice sheet that sits around two metres above sea level. However, in areas that do not melt seasonally, the ice can vary from four to twenty metres thick. The Arctic is ice bound, and in winter, there is little or no sun at all, a kind of endless night. In summer, there is almost constant daylight, as in no darkness. It is a most extraordinary biome with delicate ecosystems, and its biodiversity have adapted to these challenging conditions over many millions of years. Even the human species has adapted and made the Arctic home. Clothing, housing, hunting, and gathering all illustrate human adaptation to the extremely harsh conditions experienced by those who live in such extreme conditions.

The balance and harmony of the Arctic has a very important role to play in keeping Earth 'just right' for life. As well as holding a very large deposit of fresh water in ice, it also has a unique role of reflecting back into space ultraviolet rays from the sun that stop the Earth from getting too hot. White, as in ice, reflects sunlight, whereas darkness, such as the ocean, absorbs heat. Reflecting just the right amount of sunlight keeps the temperature in perfect balance for all creatures that call the Arctic home.

Over decades, scientists have been visiting the Arctic to do their research. What they are finding in recent times is a cause for great concern. They are busy documenting changes to the Arctic, especially the amount and quality of sea ice. Their observations are supported by satellite images of the Arctic, and what they have discovered is that Arctic sea ice is rapidly shrinking year by year. This has huge consequences for creatures that depend on the winter formation of sea ice to access their food supply. It also means a lot less ice reflecting sunlight, which, in turn, means more heating of the ice, therefore more melting.

Scientists have been able to draw out core samples of ice that tell the story of the response of ice to global warming. The core samples provide indisputable evidence that humans have been releasing too much carbon dioxide and other toxic gases into the atmosphere, thus causing the land and the ocean to heat up. Arctic ice does not have to make speeches, conduct rallies, or write to newspapers to make its plight known to us, it just melts. Satellite images show us how much sea ice has melted and predict that it could possibly disappear in the not-so-distant future, perhaps forever.

Unlike Antarctica, which is supported by a land mass, Arctic ice is floating on top of the Arctic Ocean. Recently, a group of climate scientists went on their annual Arctic research trip and discovered lakes on the surface of the ice for the first time. They also observed holes in the ice, with water streaming into vertical gullies. They placed a camera probe in the holes to discover that the opening went right down to the ocean baseline, which meant the ice was melting from underneath. This is just another example that demonstrates, through photography, how quickly conditions in the Arctic are changing and how important it is for climate scientists and us to stay ahead of what is happening around the Earth.

A significant environmental issue that seems to be demanding a lot of attention is the impact of melting Arctic ice on the Gulf Stream. The Gulf Stream helps to regulate climate and weather patterns, and it is believed to be moving slower now, so if global warming is unabated, changes in ocean currents will have consequences as a potential 'tipping point' for environmental catastrophe. For biodiversity – that is, the animals, birds, and fish that call the Arctic home – there is a dilemma. The climatic and environmental changes mean that numbers of these creatures will have to adapt or permanently migrate to new homes, which, for many, will not be possible. There is nowhere colder in the north than the Arctic. What is important for us is that we get to know Arctic ecology and the contribution it makes for a healthy Earth. We have to be ambassadors for creatures that have made the Arctic their home as they cannot speak for themselves. Thankfully, there are millions of people working together to solve environmental issues that are clearly manifest in the Arctic, but the forces are against them, and time is not on their side.

What, in a nutshell, is happening to Arctic ecology?

- The air temperature around the world is heating up, and so that is impacting on the temperature at the Arctic, which is heating up faster than lower latitudes.
- The Arctic is heating up more than twice as fast as the global temperature average. One degree of warming

globally is equal to about four degrees of warming in the Arctic.
- Ice on Greenland is melting at an unprecedented rate, which is a huge concern for scientists.
- Arctic sea ice, which grows in quantity in the winter, is now lessening and thinning each year.
- The ice is warming from above because of less ice reflecting sunlight and warming below the ice as the ocean temperature heats up. This is evident in the formation of lakes on the ice, which is a comparatively new phenomenon creating ice crevices or gullies.
- It is another one of those feedback loop scenarios. Global warming is causing Arctic ice to melt, and ice reflects the sunlight, which leads to less ice and less reflection of the sun's heat; then more dark water results in capturing more sunlight, Arctic water heats up, and more ice melts, increasing global warming.
- The melting of Arctic ice will have an influence on the Gulf Stream, which helps to regulate climate and weather patterns.
- Polar bears are entering villages now, looking for food, because in some areas, the sea ice is so thin, they cannot hunt seals from it. This is different and worth reflecting on because the plight of polar bears tells their story about the environmental crisis they are facing now and, more potently, in the future.

The Arctic biome is a major ecosystem, a long way from my backyard, and is certainly one to watch as each year, there is further research done to confirm what is going on there. A few years ago, the news broke that a Norwegian seed bank about six hundred miles from the North Pole, which was safely buried in a mountain, became exposed because of melting ice and permafrost. The seed bank was designed as a deep freeze to preserve native seeds from around the world in preparation for the worst-case scenario – that is, the failure of agriculture as a consequence of human interference with seeds and the changing climate. Climate scientists did not anticipate the thawing of permafrost and the melting of ice to be so severe or so quick. The seed vault is one hundred metres long and designed to be self-contained and self-maintained, but it now needs twenty-four-hour supervision to maintain the temperature inside the vault to preserve the native seeds. It is one of a number of seed banks, and it is sometimes referred to as the 'Noah's Ark of seeds'.

There are so many tell-tale signs about what is going on with the healthy functioning of our home planet. The interlinkage theory applies to us and the Arctic; in fact, climate scientists believe that what is happening is so serious and so visible that the Arctic is classified as one of the major 'tipping points' that can and most likely will impact on sea level rise and vital ocean currents that help to regulate climate and seasons. None of us are immune from what is happening in the Arctic.

Water Ecology

Water is an absolute, non-negotiable necessity for life. It is an infinite resource, a gift of Earth, because it is continually recycled. Earth is a water planet, and humans are water people. Let's do some pretty special maths; 70 per cent of Earth is covered in water, and our bodies are approximately 70 per cent water. Of all the water on Earth, 97 per cent is held in the ocean as salty water, while 3 per cent is fresh water; 2 per cent of the 3 per cent is held in ice at the poles and glaciers. This means that we only have 1 per cent to share among seven billion people and all plants, birds, insects, reptiles, animals, and creatures that love and need water.

Fresh water is held in rivers, wetlands, lakes, and creeks and underground as artesian water. Of all the water on Earth, not one drop is added, and not one drop is lost to the Earth; all water is recycled through nature's water filter systems – that is, soil, plants, the atmosphere, and the ocean. Water continually goes through the process of precipitation, evaporation, and condensation. This is called Earth's great water cycle. The sun heats the ocean, water vapour rises and condenses as clouds, and over time, clouds become heavy, and rainfall results.

Water takes on three forms. It can be solid (as in ice), liquid (as in a river), or gaseous (as in steam or water vapour, as in clouds). In every drop of water, there is exciting stuff happening with molecules; also, a single drop of water is full of microscopic organic life. Scientists believe from their research that life began in the ocean. That might help to explain why our tears, sweat, and urine are salty, but as tears, sweat, and urine evaporate, they leave salt behind and return to the water cycle. The average person in Australia uses many litres of water a day; given our allowance, that is about 150 litres a day. In the past, one flush of the toilet can use as much as seven litres! However, some people in the world only use about four litres a day for all their needs. To maintain a healthy body, it is suggested that people drink two litres a day, so that is a huge demand on the water cycle. Fortunately, some people are beginning to realise that water is a precious commodity, and so farmers, factories, and people in general are becoming more water conscious. All manner of water-saving devices are being invented, farmers are reviewing the way crops are irrigated, storm water is being captured, and more desalination plants are under construction as backup systems. It is the advent of a new respect for a gift of Earth that we have taken for granted for such a long time.

An anecdotal story from my experience: I had a large backyard, and so I decided I would create a beautiful garden. I paved pathways with beautiful border plants and laid in a carpet of lush lawn resplendent with an irrigation system. My brother-in-law, a farmer, came to visit, and I was showing off my magnificent garden. He shook his head and said to me, 'I thought you are an environmentalist, Marg.' 'Yes, too right,' I replied. To my response, he said, 'I don't think so. More water is poured onto suburban lawns than I have access to for growing my crops to feed people.' I thought about his appraisal of my absolutely splendid garden and decided the lawn had to go, and so up it came, and I planted native shrubs in its place – the power of education to make a difference.

Water is currently under duress. Over the last hundred years, water courses such as rivers have been a way of getting rid of waste, especially sewage and chemical waste from factories. Rubbish, through storm water runoff, also pollutes rivers and streams. Rain falls through polluted air as acid rain, so soil and water are polluted. Acid rain falls onto vegetation with serious outcomes, such as trees being poisoned by acidic rain from above. I remember reading somewhere that trees were dying on the top of a mountain range. Scientists worked out that as heavy clouds approached the mountains, they would let down their water to get over the mountain, so the trees were being poisoned from above.

For those of us who only need to turn on a tap and out flows beautiful fresh water in abundance, we have some work to do to understand that our water usage is unsustainable given rising temperatures and the potential evaporation of precious water. We must commit to the reality that our water behaviour is untenable and certainly not sustainable in the long term. Rising environmental consciousness is important if we are to restore and preserve the integrity of water for today and future generations. Water is a necessity for life. It is extremely precious, and we must defend its right to health by being water wise. Water has rights of its own to be healthy. This idea is now being supported by law in various countries. When it comes to ecological conversion, this is a big one to get our finite minds around. Are we ready to defend water's right to be protected from pollution?

What, in a nutshell, is happening to water ecology?

- Water is a gift of Earth, not a resource to be used and abused.
- Fresh water is the 'life blood' of biodiversity on land. It links every land creature together, including us.
- Fresh water is intimately linked to the ocean. As ocean water evaporates, clouds are formed, which let down their water onto the land.
- The chemistry of water, especially in the ocean, is changing because of pollutants from land use and from acidic rainfall.
- Nature's ecosystems clean water for recycling, but today ecosystems are being overwhelmed with this task.
- A multiplicity of creatures that call water home are under threat from pollution.
- People are upsetting the water balance around the world by bottling it and sending it out to a global market.
- There is water scarcity in many countries because of changes in the water cycle brought on by global warming changing the weather patterns.
- The ecosystems to which water is interlinked are changing because of practices such as deforestation. Forests help to regulate rainfall, so if they are destroyed, that changes the water cycle.
- Water is being exploited randomly as a resource to produce various products. It is used in megalitres to produce all manner of products, many of which are unsustainable; for example, the production of coal requires megalitres of water.

Water ecology is in deep trouble. Water is interlinked with all of creation. Scientists argue that human behaviour is upsetting the water balance around the world. If we take access to water for granted, as we do in some countries and Australian cities, then we are not paying attention to what is going on in the water world around us. The plain fact is that water will be available to us until it is not as Mother Nature can turn off the tap or give us flooding anytime she chooses.

Are we aware of where our water comes from, who owns the water that comes out of our taps, and is it a reliable source for the future? What we do know is that large multinational companies are very interested in owning and controlling the supply of water, even the water courses/systems in countries. Water is a good investment as it is deemed to be the next big market commodity – 'blue gold'. Water ownership is already a source of conflict for some people; as a result, water laws are already being determined and regulated by the United Nations Convention, and there are international agreements on principles to protect the right for everyone to have access to clean water as well as how to resolve conflicts over water. This is a real issue for people who depend on the flow of rivers for their water security, especially when a river flows through a number of countries or states, such as in Australia. Water ownership and water security are issues to watch unfold. However, you do have to wonder why countries and corporations are buying up water courses in countries other than their own. What do they know that we don't know?

So we do have to ask ourselves, do we value water in its own right to be healthy and available to all biodiversity? Surely, the protection of water's rights for its own sake is a cause worth fighting for. Are we joining the dots regarding the plight of water? Are we ready to be a voice for 'water rights' considering water is kin given that we are 70 per cent water?

Coral Reef Ecology

Coral reef ecosystems are absolutely vital to the health of the ocean. Reefs began forming about five hundred million years ago, so they have a very long history. There are three kinds of coral reefs: fringing reefs, which develop in shallow waters quite close to coastlines; barrier reefs, which usually develop parallel to a coastline, separated from land by a larger distance of water; and atolls, which grow around islands formed by volcanic eruptions. A coral reef begins as a single polyp, a very tiny creature that attaches itself to a firm surface. Over time, these tiny creatures live and die, and slowly, the reef builds into an underwater edifice that supports an incredible amount of sea life. Coral polyps are beautiful to look at but not pleasant to touch as they can have stinging nettle-type tentacles that come out at night to snare their food, usually plankton, as it floats past. Corals are known for their beauty, but it is not the coral itself that has bragging rights; it is the tiny creatures that live in the polyp, called zooxanthellae. It is the beauty of underwater reef wonderlands and marine life that live on the reefs that attracts scuba divers to venture into the water to be enthralled with the magnificence and mysteries of reefs.

Coral reefs provide a wonderful service to the natural world. First, they provide fish and sea creature nurseries for about a quarter of ocean fish. They support a huge biodiversity and are extremely beneficial to humans as well. Islander people depend on fish from coral reefs as their staple diet. Reefs form a natural barrier that helps to protect coastline dwellings from storms that can destroy property through flooding from higher-than-usual tides. Coral reefs also offer scientists an opportunity to seek and discover medical cures for illnesses and diseases, similar to medical benefits for humans that can be found in rainforests. They also have an intrinsic value and intrinsic rights as creatures of Earth. However, coral reefs are threatened species.

Australia's Great Barrier Reef began forming about eighteen million years ago, and it is the largest coral reef on Earth. It is situated on the east coast of Australia and is over 1,600 kilometres in length, so it can be seen from space. The Great Barrier Reef is a tourist attraction, and people visit from around the world, bringing their tourist dollars with them. Thousands of people are continually campaigning to have the Great Barrier Reef protected by law. There is an army of 'Save the Great Barrier Reef' people who campaign furiously for the protection of reefs. These people are marine scientists, environmentalist activists, and everyday people who get the importance of preserving reefs for the sake of all biodiversity that depend on them. Australians have been put on notice that as custodians of the Great Barrier Reef, they have a job to do – that is, to protect the Great Barrier Reef. This is in our backyard, so all Australians are called to arms to defend the rights of the reef.

After millions of years of growing, thriving, and enjoying life in the wild, coral reefs are now endangered by human activity. There is not an ecosystem in the world that has not been touched by humans and consequently damaged or destroyed, and reefs are no different. They are extremely sensitive ecosystems as they readily respond to changes to their environment, especially any warming of the water. Environmental marine scientists have alerted us to the fact that a quarter of the world's coral reefs have already been destroyed, and another two quarters are seriously threatened. The threats come from overfishing and thoughtless, inconsiderate tourism, especially boating. There is also the issue of pollution from urban sewage, industrial chemicals, oil spills, and coral mining as well as ocean warming because of the changing climate, which are the main culprits. Coral reefs are living icons of Earth, and importantly, we want them to be safe forever.

What, in a nutshell, is happening to coral reef ecology?

- The Great Barrier Reef and coral reefs around the world are living icons.
- The changing climate is the greatest threat to the health of reefs through global warming, which is heating up the ocean, especially the shallow water that surrounds reefs.
- Corals are extremely sensitive to any change in the climate. They cannot adapt fast enough, so scientists hold great fear for the health of reefs.
- Global warming heats the ocean, which destroys the reefs' colourful algae, and the consequence is that the corals starve and eventually die.
- Half the Great Barrier Reef has been bleached to death in the last twenty years, and its destruction has exacerbated over the last five years. Photographic evidence of what was and is now is readily available.
- Pollution is impacting on the water quality of reefs. The pollution comes from runoff from the land from storm water discharge into the ocean, debris left on beaches by careless folks, waste from factories and agriculture contaminants that find their way to the ocean, as well as oil spills, to name a few.
- Coral reefs are also threatened by ocean acidification from toxic rainfall that falls through the 410 ppm of pollutant chemicals in the atmosphere.
- Coastal development, as in housing and infrastructure, is also giving the reefs grief.
- As conditions for coral reefs change, invasive species, such as the crown-of-thorns starfish, can move in and destroy the harmony and balance of the coral reef ecosystem.
- Coral reefs are another example of wilful destruction in the name of progress for some who are more concerned about making money/saving costs. Some people think they have the right to carve pathways through the Great

Barrier Reef for their ships and cargo, with potentially dire consequences for the reef.

Coral reefs are extremely vulnerable to the changing climate, and it is well documented that heatwaves instantly affect coral reefs. How can Australians not care enough and get about cutting our toxic emissions to save the Great Barrier Reef? Half the Great Barrier Reef's coral system has been seriously affected by one degree of warming past pre-industrial levels. This loss has been described as a catastrophic ecosystem collapse because it is the largest living organism on Earth. As global emissions raise the temperature to a further one and a half degrees Celsius, it is projected that 70 per cent could be lost, and at over two degrees Celsius, the whole Great Barrier Reef could be gone. So much is at stake if the Great Barrier Reef is lost to Earth's balance and harmony. So why aren't we up in arms and demanding that not another coral reef be lost? There may be no coming back for the reefs of the world anytime soon because the changing climate is happening so fast now that there is little chance of adaptation. Do we not care that the ocean's nurseries are disappearing? Without the nurseries, what becomes of the fish that depend on them for breeding? Spare a thought for the people who live on islands, who are dependent on fish as their staple food security that is supported by coral reefs.

Scientists are trying to develop heat-resistant algae to nourish heat-resistant coral as a way of saving reefs. Does that suggest the scientists believe we are really going to let the present coral reefs go the way of extinction? Are they past thinking the coral reefs can be saved? However, it is evident that by restoring the right conditions for reefs, they can return as nature has the ability to restore and renew itself given half a chance. The world has witnessed this during COVID-19; people have experienced clear skies in some places where they had never seen clear skies for a very long time. The canals around Venice were clear, and people could see fish for the first time, and animals even made their way into the silence of cities. We have to believe it is possible that we can do what we need to do to bring back the health of the natural world, but time is of the essence to assist Mother Nature.

Ten years ago, at the Sustainability Festival in Melbourne, I went to a talk given by the then Australian chief marine scientist Dr Charlie Veron. He began his career studying coral reefs in 1972 and dived on reefs for six thousand hours in his career. He confessed that he was previously a bit of a climate change sceptic before witnessing coral bleaching. The only thing I remember from his talk was that after researching the Great Barrier Reef for four decades, he believed the Great Barrier Reef only had about ten years to live if we continued along the then present path. Recently, I heard him speak again, and his message had not changed except it was more urgent, more devastating in that the loss of coral reefs is thought to be as much as 50 per cent. He argued that marine ecological collapse will happen as a third of marine species begin their lives in coral reefs. While listening to him, I could feel his pain and despair. This is a man who lives deep ecology. He feels the loss of corals in his heart and is deeply affected by it just as he would feel in losing a family member. If we do not learn from the lived experience of someone like Dr Veron, then how will we ever make the changes so vital to our survival as a species?

Deep ecology is knowing that we are all, the whole community of beings, deeply intertwined in nature, and there is no escaping it because we are kin to all of nature. A deep ecologist, a person who absolutely understands our connectedness to the natural world, would say, 'I am defending coral reefs because I am coral reefs.' It is so hard to get our feeble minds around deep ecology and the need to be in empathy with the plight of biodiversity such as coral reefs.

Wetlands Ecology

Just like Cinderella soil, we can think, *Wetlands – what are they?* You might ask yourself if anyone has ever raised a conversation about wetlands with you. Wetland ecology has a vital role to play in the health of Earth, but who knows that important fact? They actually cover a significant amount of Earth's surface, so it is very important to understand what they do to keep Earth in a natural state of balance and harmony.

You might know wetlands as swamps, marshes, or bogs, and they can be freshwater or saltwater. Some wetlands have brackish water, which means the water is stagnant and salty, and whilst they are not perhaps a pleasant sight to humans, many creatures are quite at home in them. Mangrove wetlands, for example, are found along the coastline, while marsh wetlands form along waterways such as rivers, creeks, and lakes.

Flooding is a characteristic of wetlands. Tidal wetlands are refreshed from ocean tides, whereas non-tidal wetlands depend on rain or underground water seeping up through the soil. In the case of peat bogs, it is more about melting permafrost, ice, or snow. Wetlands, as the name suggests, are where water naturally forms in low-lying areas of land and remains there permanently as a general rule, although there are some seasonal wetlands. As an ecosystem, wetlands have their unique biodiversity of plants, insects, reptiles, fish, and birds. Mosquitoes and frogs are synonymous with wetlands – you might say 'frog and mosquito paradise'. Turtles, tortoises, and snakes as well as many types of fish also abound in wetlands.

We know that every ecosystem makes an incredible contribution to Earth's health, and wetlands are no exception; they are busy about many things. Wetlands assist other ecosystems in purifying water by removing toxic pollutants. They have a huge role in flood and erosion control and limiting storm damage; tidal wetlands provide a buffer between land and sea. Freshwater wetlands provide a reservoir of water for the irrigation of agriculture as well as water for livestock. Some wetlands support forests for timber and nurseries for fish. Importantly, wetlands recharge groundwater and are carbon dioxide sinks. Carbon sinks hold carbon dioxide that is absorbed from the atmosphere. The wetlands' capacity to sequestrate carbon dioxide is thought to be greater than the mighty Amazon Rainforest, so they are considered to be the 'kidneys of Earth'. Strange then, isn't it, why we hear so much about the Amazon Rainforest and so little about wetlands? Are the 'lungs' more important than the 'kidneys'? Maybe we should/could focus more on the Cinderella ecosystems as they seem pretty important to our survival.

Thankfully, there are people who are drawn to wetlands as scientists for research or tourists for boating, fishing, birdwatching, and camping. Where housing estates are built on reclaimed wetlands, the wetlands provide beautiful parks and gardens and offer a wonderful experience of wildlife for those who take time to enjoy them. However, wetlands are experiencing the impact of human activity, and many concerned people are raising alarm bells about the need to protect and preserve wetlands. Destroying wetlands by filling them in to create housing estates, overfishing through aquaculture, the introduction of invasive species, and pollution through agriculture runoff containing fertilisers and pesticide chemicals are just some of the issues challenging wetlands today. Wetlands are not iconic biomes like rainforests or oceans, so what is desperately needed is education on wetlands' irreplaceable contribution to a healthy Earth. Environmentalists are extremely conscious of the need to be a voice for the voiceless wetlands so the creatures that call wetlands home can have their habitats restored and protected for them.

What, in a nutshell, is happening to wetlands ecology?

- Humans are the biggest threat to wetlands. We humans are responsible for wetland destruction and degradation because we don't consider that wetlands have a huge role in keeping Planet Earth healthy.
- People drain wetlands for urban development; they dam up wetlands to create lakes or ponds, they put infrastructure on, in, and around wetlands, they divert water flow in and out of wetlands, and in general, they upset the hydrological cycle or conditions for creatures that live in the wetlands.
- Wetlands can be subject to having to deal with industrial waste and municipal sewage as well as runoff from urban and agricultural activities, which, of course, pollutes the wetlands and destroys biodiversity.
- As integral to marine ecosystems, wetlands also suffer from the loss of habitats for their biodiversity of creatures that call wetlands home.
- According to Greenpeace – a worldwide environmental movement that is out there, checking the environmental status of Earth – 2.3 million hectares of the wetlands in Brazil, home to unique biodiversity. have been burned in very recent times.

To recapitulate, we do not think about wetlands very much, that kind of Cinderella ecosystem, as having a vital role to play, but wetlands are considered the 'kidneys' of Earth. A human with kidney issues is in trouble, so wetlands as the kidneys of Earth are experiencing an unhealthy situation. The traditional role of wetlands is to purify water as it passes through them. Again, with more potent pollutants entering the hydrological system, the wetland kidneys are under stress. Only education about the role that wetlands have in keeping Earth's ecosystems in balance and harmony will change the way we think about wetlands and, consequently, the way we value and want to protect them. Do you have any wetlands in your backyard that might need some loving care?

All wetlands are amazing in that they just keep on doing their job, a job we are neither familiar with nor, in general, grateful for, but we could get to know what wetlands do to help maintain a healthy Earth. Here, it is appropriate to say that we must get to know how Earth functions in unison. If we get to know about the needs of Mother Earth and all of her children's ecosystems, then we can learn to love her because thankfully, we take good care of that which we love.

River Ecology

Rivers are life-giving everywhere they go. They are sometimes referred to as the 'veins of Earth' as they take life-giving water across countries and continents. Villages, towns, and cities are established on river systems because access to water security is paramount for the survival of all who live there. Rivers have length, depth, width, and (strangely) time. They rush in their beginnings and meander further downstream. Rivers are busy creatures. They provide water for a biodiversity of creatures, especially an abundance of wildlife, as permanent residents or visitors. Around the world, rivers are also used by humans for bathing, washing, transport, irrigation, recreation, spirituality, and getting rid of waste. However, many mighty and majestic rivers no longer reach the sea. This is a cause for alarm.

Our traditional approach to rivers is not one to be proud of, really. We have been arrogant in the way we have treated rivers, totally lacking in respect for their health. Humans have exploited rivers by creating dams to supply towns and water for industries and agriculture. To add insult to injury, humans then return polluted wastewater and sewage into the river systems to get rid of it, and rivers generally find their way to the sea and take debris with them.

The Murray–Darling Basin, resplendent with many fine rivers, has a very long history that goes back some forty-plus million years. It is an amazing river system that shares its gifts with Queensland, New South Wales, Victoria, and South Australia. The river system encompasses about 14 per cent of Australia. Rainfall on the Great Dividing Range flows westward inland until it finally gathers in the Murray River and flows out to sea near Adelaide in South Australia. The Darling River is the longest river in Australia and flows through the Murray–Darling Basin, which is sometimes called the 'food bowl' of Australia.

However, rivers in the Murray–Darling Basin are not in good ecological condition. Monumental efforts are being made now to restore and preserve this magnificent water system. Environmental flows are now part of the conservation program as it is recognised that the river system is a living ecosystem and has its own needs to remain healthy and productive. One apparent problem with environmental flows is that those who turn on the tap do not usually consult with Mother Nature, and flows are not always directed in time or volume that works with nature. Environmental flows are a source of dispute along the Murray–Darling Basin, and there is possibly some validity in the debate as even environmental flows are an example of human dominance over the river system by dictating when and where the environmental flow should be released.

What is also acknowledged is that if the Murray–Darling Basin river system is not cared for, then it will not produce the food needed to feed millions of people. The Murray River has, in recent times, not reached the sea. This can mean that too much water has been withdrawn, causing salinity to build up in the river. A consequence of this reality is a threat to wetlands, birds, animals, and fish that depend on a healthy river.

This account of the Murray–Darling Basin is synonymous with other river systems in the world. Wherever people live, it is important that they get to know their river, who owns it, who cares for it, and how to nurture it. Taking care of the natural world, which includes rivers, is everyone's responsibility. This is crucial for water security in every country as there is the probability or at least the possibility of wars over water security.

What, in a nutshell, is happening to river ecology?

- Rivers find their natural paths to the ocean, but humans have dammed them up, diverted them, and drawn out so much water that they often do not reach the sea.
- Water levels in many rivers are decreasing to the point when they may no longer be able to service the people who depend on them.
- The overexploitation of rivers is an issue as the flow of rivers is unnaturally restricted when there is less volume to push water through its natural course.
- Many rivers are polluted from agricultural runoff, industrial waste, and, of more recent times, flooding. Storms create the flooding of towns and infrastructure, so many potential pollutants are washed into rivers in their wake.
- Plastic is a major pollution problem for rivers; many rivers in the world are contaminated with plastic debris.
- Pollution in rivers creates monumental stress for biodiversity that dwell in them or live by them. The pollution collectively puts thousands of aquatic lives at risk.
- Rivers are subject to global warming as evaporation is an issue, especially in rivers that do not flow all year round.

- Free-range livestock can pollute rivers and destroy the banks, which are habitats for wildlife.
- Salmon are dying en route to their spawning grounds in some rivers because of the warming of the rivers.
- Rivers are a vital source of water for many millions of people, so if they dry up because of the lack of traditional rainfall, evaporation from global warming, or less ice on mountains, then people may become refugees in their own countries.
- Water is 'blue gold', and the prediction is that in the future, there will be wars over water, so who owns and protects the river systems of the world is not only an environmental issue; water security is a social, geographical, political, and economic issue.

What is important is, given that the river systems are the 'veins of the Earth', we need to know the rivers that our lives depend on. Not only are water issues on the rise as the demand for more and more water for industry and agriculture grows, but also, there is the expectation of population growth around the world. These are serious considerations, so the challenge is for you to get to know your river system, its history, its environmental status, where it begins, and where it flows.

As for the fish that call rivers their home, they are now in serious decline because of drought, invasive species, and pollution. When we were children, our dad used to take us out to the local creek to fish, and we would catch enough fish in an hour to supplement the family diet, but in that same creek today, we could fish all day and catch nothing – no edible fish there now to be caught, only an invasive species called carp, which has destroyed native fish ecology. This, I imagine, is what is happening to ocean fish. We will not notice the fish are gone until there are none to catch.

It is important to understand human dependence on healthy river systems. Most rivers begin with rainfall in the mountains or elevated land, but rivers that flow because of melting ice or snow are a particular concern because with global warming, the ice melt may be reduced or, in some cases, disappear, so the river systems in some countries could be seriously at risk. It is our task to educate ourselves as to where and how rivers form and add in the unreliability of rainfall because of global warming, seasonal variations, and deforestation. Forests bring rain; that is why they are called rainforests. If rainforests are cut down, where will the river rain come from? The signs of the environmental degradation of rivers are all around us if we choose to see them.

Glacier Ecology

Who thinks about glaciers? We do not live on glaciers, so we don't really hear much about them. When you do think about glaciers, they fire the imagination; they have a certain aura about them. Glaciers have an ageless wisdom and ascetic beauty that draws people to them for a visit or photo opportunity. Glaciers are huge, strong, steady and move, albeit at a very slow pace. Australians are not familiar with glaciers, probably because Australia has a generally hot climate. The highest peak is Mount Kosciuszko in the Snowy Mountain National Park. It rises over two thousand metres above sea level and is snow covered in the winter but mostly ice free in the summer, so glaciers do not form on Australian mountain ranges.

Glaciers form when snow remains all year round; over time, the snow turns to ice, which compresses and builds up with every snowfall. The snow actually becomes solidified, forming glaciers. Glaciers vary in size, from small, the length of your street, to very large and extremely huge – that is, many kilometres long and very deep. Glaciers can be rocky or amazingly beautiful, like the Franz

Josef Glacier in New Zealand, which has a radiant white/aqua/pale blue colour that fascinates visitors. This glacier is a huge tourist attraction.

Glaciers formed over millions of years. They are referred to as 'the Third Pole' as they hold so much fresh water in their ice. Some of the largest glaciers are remnants of the last Ice Age, so they have a very long story to tell about their lived experience. Believe it or not, they occupy a significant part of Earth's landmass; although most glaciers are found in Antarctica and Greenland in the Arctic Circle, over forty countries have glaciers. Glaciers provide an amazing service to people, flora, and fauna as the meltwater flows through various countries. The Himalayan Mountains are home to thousands of glaciers, and their meltwater is vital to many millions of people in Asia who depend on the meltwater for drinking, domestic use, agriculture, and power generation.

Glacial ice holds most of Earth's fresh water if you take every glacier into consideration. Antarctica's ice sheet alone is a glacier, and it holds enough water that if it were to melt, the sea level would rise by many metres. For some environmental scientists, glaciers are of special interest. Since 1980, changes in glaciers, such as 'retreating' because they are melting at an unusual rate, have been important signals regarding the changing climate. A retreating glacier means that the mouth or terminus of the glacier is further up the mountain than previous years. As well as glaciers retreating, there are examples of glaciers disappearing, which is catastrophic for those who depend on them.

The warming global climate is thought to explain what is happening to glaciers. As global warming continues, no doubt glaciers will tell this story in the speed of their ice melt. Glacial ice does not make speeches about its condition; it, quite simply, just melts. Glacial melting or thawing is one of the most obvious signs that the climate is changing. There are plenty of photographs that clearly illustrate when there was a glacier and that the glacier has disappeared. The Himalayan Mountains glaciers' melt loss has doubled since 2000. This is an alarming statistic and does not augur well for the future of meltwater so important to feeding major rivers in China and India, for example.

Every ecosystem supports another; glaciers are linked to rivers and rivers to food bowls to feed millions of people. Earth science, therefore, is very important because the more we know about our home planet, the more we will understand the way it functions and the more we will love it and want to preserve it. Glaciers have an extremely important role – that is, to capture snow, store it as ice, and release it as meltwater in the summer, thus filling rivers for all that depend on them to survive.

What, in a nutshell, is happening to glacier ecology?

- If Antarctica can be referred to as the 'big meltdown', then glacial scientists refer to what is happening to the glaciers as the 'big thaw'.
- Glacial ice is subject to a higher intensity of global warming, in the order of three to three and a half degrees Celsius higher than on lower lands.
- Glaciers are very sensitive to global warming. Apparently, only a one-degree-Celsius rise in temperature can have a huge impact on glacial ice.
- Glacial lakes are forming which is a comparatively new phenomenon. That is very different and potentially catastrophic for the people who live below the glaciers. If the lakes burst, then devastating flooding can result.
- Glacial ice loss on the Himalayan Mountains is accelerating, which is a threat to many millions of people in Asia who depend on glaciers for their water security. The ice melt produces water for several mighty rivers that provide

water for irrigation and hydroelectricity, especially in Asia, particularly China and India.
- In the past forty years, some mountains have lost something like a quarter of their ice. As the atmosphere heats up, glacial melt is expected to accelerate.
- The big question for scientists is as the climate warms, how quickly will the glaciers melt, and what would be the environmental consequences if they do melt?
- Glaciers hold a huge amount of fresh water, and if they continue to melt, which is what scientists fear, the meltwater can change the chemistry of the ocean by adding fresh water, so the ocean would be less salty.

As a normal busy person, I have not given a thought to glaciers. We don't know much about the role glaciers play in the health of the Earth. However, the scene is set by the scientists who have followed their demise for over forty years. Now there are satellite images that confirm their predictions that with the influence of global warming, the glaciers are melting and the melt is accelerating. Again, it is a case of wondering if we can get global warming under control and, in time, stop it, then an environmental catastrophe can be avoided.

Dare we begin to think of the consequences to human and, indeed, all life if worldwide glaciers were to drastically decline? The scientific prediction is that in less than eighty years, at the present rate of global warming, the Himalayan glaciers could completely disappear. This, of course, would be absolutely catastrophic for millions of people. However, this prediction does not consider what might happen if global warming were to accelerate out of control because of the pollutants that are already in the atmosphere.

It is clearly a case of environmental literacy – that is, reading the signs of the Earth. For us who are protected by where we live and how we live at present, we must pay attention to what is happening in the real hotspots of the world that confirm evidence of climatic change. There is probably no clearer sign that human activity is changing the chemistry of Earth than following the story of ice. As I write this, I am thinking, *No wonder we don't listen to the scientists – because they are not bearers of good news in general.* It is so hard for them, as people first, to share the bad news, but that is their job. In their early careers, they would have studied the wonder and awesomeness of glaciers, how they form, how they contribute to a healthy Earth, glacial biodiversity, and all the lovely things about glaciers; now they are reduced to being bearers of what is happening to their awesome glaciers. Spare a kind thought for those many climate scientists who are sadly bearers of bad news.

I remember watching a video one time about a family that visited Nepal. They met up with an old glacier guide in his eighties who took them to visit a retreating glacier. The family saw the ladders that people had climbed down to the glacier terminus over the decades – that is, successive ladders following the retreat of the glacier. What I remember most was the old guide, who lamented the passing of the glacier's depth and width. He was deeply saddened by the loss and fearful for the future of his treasured glacier. Again, there is no room for doubt here as the retreating glaciers are visible to the eye and photography.

Mangrove Ecology

Mangroves are described in the dictionary as tropical trees that have exposed roots and grow along the coast at the water's edge. Mangrove ecology is another 'Cinderella' to the great ecologies, such as the ocean or forests, but life would not be the same without mangrove ecosystems. Mangrove ecology studies the interactions of its biodiversity, which includes land

and water creatures. Mangrove habitats are located generally along sheltered coastlines and usually associated with the tropics as they thrive in warmer waters; however, other plant species far removed from the tropics perform similar services to the mangroves. Mangroves can also creep into inlets such as estuaries or lagoons. They are sometimes the fringe dwellers of wetlands and forests.

Mangroves are important not only for the biodiversity that they support but also the role they play in protecting coastlines. They provide a natural barrier between the surging surf of seas and the land that they protect. They have an amazing ability to rebound from natural challenges such as high winds and buffeting waves. Mangroves also have a role in maintaining water quality as they capture sediments and excessive organic or inorganic nutrients in the water.

Many different creatures live in mangroves above and below the water. Plant, insect, bird, and aquatic life that call mangroves home are also adapted to tidal surges and tough weather conditions. Because their plant life is dense and mangrove roots are in the water, they also provide a safe nursery for marine life, especially very small fish and members of the reptilian family that are semi-aquatic. Many varieties of marine life, such as shrimp, spawn their young amongst mangroves' roots; therefore, they help to support and secure commercial fishing.

Mangroves provide products as well as services, such as fish, crabs, and shrimp that find their way onto our menus. They also provide wood for fires, furniture, boats, and even housing for some people – a gift of mangroves. Mangrove ecotourism is a growing industry as some people seem to want to explore the exotic in nature. However, mangroves are feeling the influence of intense human activity. Pollution is an issue for mangroves, such as oil spills, chemical runoff, and the discharge of sewage and rubbish that find their way into mangrove habitats. They are also challenged by the changing climate and needing to adapt to rising sea levels and warming water, which, of course, affects all coastal mangrove ecosystems. Increased sedimentation is also having a negative impact by upsetting the balance and harmony of life for mangroves. Any changes are not welcome and are not quickly adjusted to because adaptation is, for most ecosystems, a very long process; mangrove ecology is extremely fragile. Although mangrove swamps are a bit scary for people, it is still in our best interest to learn about and appreciate their unique role.

What, in a nutshell, is happening in mangrove ecology?

- Mangroves straddle that rare space where land and sea meet. Their branches and foliage support reptiles and birds, while their roots, submerged in salt water, protect tiny fish and other marine creatures.
- Mangroves are not usually credited with the work they do for the balance and harmony of ecosystems, such as the sequestration of about four times more carbon in the atmosphere than trees. Mangroves are powerhouses for carbon capture and sequestration. They excel at storing carbon, but when they are cut down, they release the stored carbon. This is another domino effect.
- Mangroves provide a carbon sink, which means sequestering pollutants in the atmosphere, and also a pollution sink, which deals with chemical, industrial, and human waste.
- Like all other ecosystems, mangroves are subject to change brought about by anthropogenic activity, such as burning fossil fuels. They are subject to climate extremes, just like the glaciers and coral reefs.
- It is estimated that at least 35 per cent of mangroves have disappeared in the world because of cutting them down or changes to land use.

- The impact on biodiversity in mangroves is manifest in the loss of habitats, pollution (especially chemical pollution), tropical cyclones that now present with greater velocity, and high or low sea level events, which also influence their health or recovery.
- In some countries where the mortality of mangroves is clearly visible and their loss is noticeable because of the ecological job they no longer do, replanting projects are underway in the hope that they can be restored, but that will be a huge challenge as environmental conditions that grew them in the first place have changed.

We may not think that mangroves are very glamorous creatures in nature, but one thing for sure is that the human species cannot do the work of mangroves. My one experience of mangroves was to be attacked by swarms of mosquitoes; I knew that mixing with mangroves was definitely not a place for the human species and so best left to the birds, reptiles, and fish that feed on the mossies. Every ecosystem has a job to do that is vital for the stability and coherence of Earth. Environmental scientists and people who have been affected by their disappearance are trying to re-establish lost mangroves, and to do that important work, they have to understand wave energy, the acceptable degree of salinity, soil composition and water pH, sediment composition, nutrient concentrations, wind velocity, rising temperatures – the list goes on. Mangroves had to adapt to all of the above to provide their special service to the biodiversity they support. Ecosystems such as mangroves formed over millions of years and slowly adapted to a habitat that supports them. So for us before we start destroying anything in nature, we really need to do our homework as to whether short-term gain is worth long-term pain, like flooding.

Artesian Ecology

Underground water is life-saving for many millions of people. On a daily basis, many people draw water from the Earth to supply all their needs. Australia may not have the highest mountain or the largest lakes, but it does have the largest artesian water catchment in the world, called the Great Artesian Basin. The Great Artesian Basin is a reservoir of fresh water that lies under the north-eastern states of Australia. Artesian water is the only reliable water supply for inland dwellers, especially First Nations people of Australia who have depended on it for more than forty thousand years. When new settlers came to Australia, the discovery of the Great Artesian Basin opened the Outback to explorers and farmers as it provided water security for domestic use as well as livestock and irrigation.

Words associated with artesian ecology are 'discharge' and 'recharge'. Discharge occurs naturally through springs – that is, water seeps up through the ground, sometimes creating a wetland or oasis. Water is also discharged through bores, which are holes drilled down into the artesian reservoir, a pipe is pushed down to meet the underground water. Water pressure pushes the water to the surface in most cases, although water can also be pumped to the surface with the help of a windmill. Other people dig wells and brick them up with stones to draw up underground water. In deserts and arid landscapes, this method of securing water is the norm. Recharge is artesian water reservoirs renewing themselves. This occurs through rainfall seeping down through layers of permeable or porous soil and sandstone.

There are two types of artesian reservoirs. They are 'replenishable' and 'non-replenishable'. Non-replenishable reservoirs are finite – that is, no more water is flowing into them, and so water tables in some countries are being depleted of water. 'Replenishable' means that the water tables are being continually topped up through rainfall. What is happening is that many artesian aquifers or reservoirs are drying up because too much water is being drawn out of them, and many water wells have already dried up. This creates a very serious problem for people who depend on them for their water needs, especially in those places that experience the lack of rainfall or are in constant drought. In some countries, agriculture, which once depended on artesian water, has been abandoned because of the over-pumping of artesian water. Water above or below the ground is a very precious gift of Earth. The consequence of the loss of artesian water means that people have to move off their land and settle in another place where water is available.

What we know about artesian ecology is that they are very fragile ecosystems and need to be respected and carefully managed with no waste. As the population grows in various parts of the world, so does the need for safe fresh water. Farmers and villagers are more aware now of the need to make every drop count and not take access to water for granted. We can learn from the experiences of other countries that some artesian reservoirs are finite.

What, in a nutshell, is happening in artesian water ecology?

- Many artesian reservoirs have been used and abused. Little did people know that the well could run dry!
- Artesian groundwater is also a victim of fertilisers, pesticides, and herbicides as they can soak into the ground and into the groundwater. Those 'cides' seem to give a lot of creatures grief, but thankfully, there are policies up and running to eliminate them.
- Manufacturing and mining are also culprits regarding the pollution of groundwater. This is why there is a huge environmental movement campaigning against coal mining and fracking for gas.

- Artesian water is the lifeblood for many farmers. The extreme use of artesian water can impact on geographical and geological activities.
- The loss of artesian water has implications for the ecological environment and the biodiversity it supports – that is, vegetation, animals, and birds.

I guess if you take something out, then there is a void, and something has to fill that void, so collapse is also an option. In many cases of artesian bores which formed millions of years ago, they are not replenishable, so while many people around the world still depend on artesian water, if they were to fail, then the people cannot live without water; thus, it is just another case of the possibility of environmental refugees within their own countries. In the 2019–2020 drought in the eastern states of Australia, there were a number of towns without water. They had to get water trucked into their towns for months so the people could stay in their home. Artesian water, wherever it is found, is so unbelievably valuable, but we have yet to realise it and commit to protecting it.

This is only a brief summary – the tip of the iceberg, as they say – of the many ecosystems that support our lives. In our ignorance, we, in general, have not taken the time to understand how ecosystems operate as a collective, how they function daily to keep the balance and harmony of the conditions that made life possible in the first place and continue to support life. We tend to go about our days using whatever resources, gifts of Earth, that we need without thinking that we could possibly change and eventually destroy those ecosystems so vital to a healthy Earth. Earth is alive, it lives, and it has its own needs to support all life. It is vital to understand the ecology of Earth to appreciate the impact of our actions, not only on the human species but also on all biodiversity. I keep on mentioning that ecosystems are all interlinked and they depend on one another to keep a balanced and harmonious biosphere. All life, including our lives, depends on all ecosystems continuing the services they provide for a stable, liveable Earth. As an intelligent species, we must understand that the human species cannot survive without the ecosystems that support our lives. That is a non-negotiable fact. After reflecting on the major ecosystems, we can turn our minds to the role biodiversity has in contributing to healthy ecosystems and how they are responding to the Anthropocene Era.

CHAPTER SIX

What, in a Nutshell, Is Happening to Biodiversity?

We can now turn our attention to worldwide biodiversity. As 'business-as-usual' people, we frequently have no idea what the multitude of biodiversity does for the health of the planet, but our ignorance is not an excuse for not appreciating the services they provide to sustain a healthy Earth. As half of the world's population live in cities, the loss of biodiversity has been creeping up on us without our knowing.

Over half of the Earth's biodiversity has disappeared over the last forty years, and we are not, in general, aware of that fact. I woke up the other day thinking an absolutely crazy thought: *When did we, as a community, stop using small envelopes?* The incoming and outgoing mail are now all business-size envelopes. Why is that crazy? It reminded me of our biodiversity that could be here today and disappear tomorrow, and we would not realise they are gone until we ask the question 'What happened to those flocks of sparrows that used to be in their thousands?' Hmmm... don't know, just disappeared!

In the first place, all creatures are members of biomes and ecosystems. Every creature enriches their biome and has a role to play in the balance and harmony of their particular ecosystem. The biodiversity today are the creatures that have shared our evolutionary journey. They represent the survival-of-the-fittest creatures that have evolved with us. They are the creatures that have adapted to the conditions of life over millions of years.

Sadly, many are now at risk from the human species as they are numbered amongst the threatened and endangered, and as many as (50 per cent) have gone extinct on our watch. We do not have much to celebrate here as the dominant species on Earth. It will be to our eternal shame if another species goes extinct. When flora or fauna go extinct, they are gone forever, and we are the poorer for their loss. So if you can bear it, read along with me, and together, we can examine our consciences.

Plant Biodiversity

Plant ecology has one of the longest histories, beginning about four billion years ago, when cells evolved to photosynthesise. This process changed Earth's atmosphere and made the herbivorous creature's life possible. About 460 million years ago, algae and fungi left the ocean for land life, and the first plants, such as mosses, began a very long story as land dwellers. Most food chains begin with plants or algae; this is because plants and algae produce their own food from carbon dioxide and water using the energy of sunlight. Chlorophyll, the green substance in plants, enables the photosynthesis process to absorb energy from sunlight, and this energy is used to make food from carbon dioxide and water.

Photosynthesis, therefore, is a process when plants take in carbon dioxide and water and, with energy from the sun, make food for themselves and then release oxygen into the atmosphere. How awesome are trees, flowers, and all kinds of vegetation that are all busy providing food for the total Earth community as well as helping to clean air and water? As well as energy from the sun, plants also need carbon, oxygen, vital nutrients (such as phosphorus and nitrogen), mineral-rich soil, and, of course, many millions of tiny and microscopic organisms to break down the minerals, decaying plants and animals. These tiny creatures also help to aerate soil to sustain plants. No human can do this extremely vital work.

Over time, plants diversified into a myriad of different species. Plants took over the land and adapted to the climatic conditions wherever the wind blew their seeds. As plants adapted to new and varied climates and terrain, so too did birds and animals that depended on them adapt to their new food source. This is one reason why there are so many different species of insects, reptiles, birds, and animals. Plants provide food and habitats for ants and elephants. Creatures that eat plants are called herbivores, and animals that eat herbivores are called carnivores. The human species is both herbivore and carnivore. Many insects, birds, and animals call plants their home, from desert cacti to rainforest vegetation, with such plants providing food and habitat for millions of other creatures we do not know. In many cases, they mutually assist one another to live and thrive. Every creature, flora or fauna, works together in a symbiotic relationship for the health of the ecosystem they call home.

Plants also have an aesthetic influence on humans as they add colour and beauty to the landscape. Have you ever wondered at the magnificent symmetry of a flower? What is important is that we value plants for their own sake as well as the awesome contribution they make to life on Earth. We can be attentive to the needs of plants and be a voice for them when needed because humans could never do the work of plants in cleaning air, soil, and water. Eco-literacy helps us to respect our interdependence with plants. Plants have plant rights. For example, weeds are plants that humans do not value.

What, in a nutshell, is happening to plant ecology?

- We take plants for granted; we do not, in general, think about their contribution to our lives or the health of Earth.
- Plants provide food and habitats for many millions/billions of creatures.
- Humans are selective about the plants they want to thrive.
- Plants have adjusted/adapted to specific climatic conditions, and this is evidenced in their leaves and flowers, such as needle-like pine tree leaves to cope with ice and snow and cacti that exist in extremely hot climates.
- Plants, in general, are territorial, so migration is not their thing.
- The four major considerations for the viability of plants are light and energy from the sun, suitable soil, a stable temperature, and reliable water, as in rainfall.
- Global warming and the human domination of plants are putting all four conditions in the variable bucket. The changing climate is threatening many plants that cannot migrate to a more conducive climate.
- Human activity is changing the chemistry of the air, soil, and water and heating up the atmosphere; it follows logically that plant life, in its many forms, is struggling to flourish or even survive.
- Once humans realise what is happening to the plants in their environs, action is usually taken to rectify the problem if it is possible to do so. A good example of this response is airlines that generate huge amounts of pollutants in the atmosphere. They have within their sign-in process an opportunity for passengers to contribute towards carbon offsets for taking the flight. These carbon offsets are very

welcome to people who are committed to replanting native forests.
- Native plants are also subject to human interference with cropping, along with the use of chemicals designed to kill any plant that is not part of their monocrop.
- Plants that have taken in carbon forever are now stressed by too much carbon dioxide to process. Plants are no different to us in the sense that we can eat a couple of pieces of pizza, but if we eat the whole pizza, we do not feel so good, and our stomachs let us know.
- Because we are the dominant species, we decide which plants live or die. Some people just love to go on a murdering spree with weedkillers. Each plant we kill affects the microorganisms in the soil that are supporting plant life. We tend to kill plants without conscience.
- In our ignorance, we do not know what each plant contributes to the health of Earth, but each plant has evolved within the balance and harmony of evolution and provides a service to some other creature.
- Each plant is precious to Mother Earth; she has no favourites.

Plants, like air and water, provide an irreplaceable foundation for life. We need to learn about the plants that share our environs and do what is needed to be done to protect native species. Flora scientists and environmentalists are extremely anxious about the loss of diversity within plant families. There are a number of seed vaults being built around the world to store native plant seeds in the event of catastrophic global warming and crop failure. These seed vaults are to preserve native seeds for posterity. I think the fact that seed vaults are being built and filled with native seeds is a good indicator that scientists think plants will be seriously affected by the changing climate. What other conclusion could we possibly draw to give us comfort? What plants are in your backyard that you are responsible for? They know you are present. Talk to them; they are kin.

Insect Biodiversity

Children love bugs. Adults tend to spray them. Actually, humans are 'big bugs' in the evolutionary story. Too much insecticide would kill us as well. Fossils of insects have been found that date insects as having evolved about four hundred million years ago, so they have been around forever. When you look closely at insects, they seem so fragile, and you wonder how they have survived all the hazardous conditions that nature throws at them. Entomologists study insects, and they are completely captivated by their awesome attributes and what they contribute to a healthy Earth. Also, there is nervous tension as the loss of insects is only now being fully assessed; the findings are startling as recent studies on the plight of insects have set off alarm bells to the point where we could be facing an insect Armageddon.

Insects were the first creatures to take to flight. Today many insects have preserved the amazing evolutionary gift of flight, which has probably been a great asset in their survival. However, other insects do not fly but are gifted in other ways – for example, panoramic 360 vision. Insects are also amazing architects as they can construct extremely complex nests. We could not make their nests for them, but we have learned a few things about construction, technology, and building from them. It is called insect mimicry. For example, a helicopter is based on the proficiencies of dragonflies. How can mosquitoes bite us and we don't know it is happening? It is something to do with the formation and action of their proboscis, so scientists are studying their special technique, and human injection needles are being modelled on the mosquito genius of impregnating our skin; however, the mosquito biting technique has not been established yet! Insect ecology, therefore, is a study of insects and their relationships with the creatures that share their ecosystems. Insects have a vital role to play in ecosystems. It has been said that if all the insects in the world and all the animals were put on a set of scales, then the combined insect community would outweigh the animals. Given such huge numbers, it is not surprising that they must be busy about many things.

Insects are significant contributors to biodiversity. They have important roles to play in decomposition and nutrient recycling through breaking down dung, tree debris, and decaying animals; for example, the maggots of a blowfly do a great job on decaying animals. Many birds, bats, reptiles, and spiders eat insects, so they are important in the food chain. Insects also live on hosts, so they are called parasites; however, sometimes it is for their mutual benefit, so we can't be too tough on them. Insects are also engaged in the all-important work of pollinating 75 per cent of flowering plants.

Insects also compete with humans because they are serious eaters. Swarms of locusts or grasshoppers can consume an entire crop in no time. Insects can devour whole plants and even destroy trees, and we know insects are also known for spreading disease. Fleas and mosquitoes, to name two culprits, are well documented as carriers of disease. For example, fleas are accredited with the notorious, historical Black Death in Europe, while mosquitoes contribute to malaria, which can be deadly. However, debates would probably conclude that regardless of the negative impacts of insects, their benefits to keeping Earth healthy and functioning to ensure harmony and balance are overwhelming. We, in general, simply do not know all the good stuff they do for us in our business-as-usual lives.

Environmentally, insects are being seriously challenged, and the biomass of insects has critically declined in numbers of species. It is estimated that 50 per cent have been lost to the Earth community over the last few decades. Entomologists fear that the loss of insects may not be noticed until it is too late as their extinction rate is far higher than those of mammals, birds, or reptiles. However, since all living creatures are interdependent and interlinked, the loss of insects could cause a cascade of bird, fish, and reptile loss as insects are the food source for many of them. Other insects are struggling to relocate to a more conducive habitat because of a warming Earth. Also, traditional habitats for insects such as wetlands and mangroves are disappearing. More importantly, the majority of insects are found in tropical rainforests, so with the daily destruction of rainforests, insects are extremely vulnerable.

We cannot and must not underestimate the severity of the loss of insects, and we may not know they have disappeared, like the small envelopes, until it is too late. Let us learn to love insects for the gift they are before we notice that they are missing from our lives and their disappearance becomes evident because we 'big bugs' can't do what they do. By the way, you probably know that insects are being farmed now in the event that traditional protein sources do not survive global warming. Members of the human species have always eaten insects anyway.

What, in a nutshell, is happening to insect ecology?

- Insects have shared our evolutionary journey, so that makes them pretty special. Some of us think of insects as pests, and we have invented bountiful ways to kill them, but really, do we know what each insect does to support harmony and balance in ecosystems? Mostly, we don't. We take them for granted and dismiss them as pretty much nuisance value. However, in our ignorance, we may be bringing about their demise and, consequently, our demise. Well, now that sounds pretty ridiculous, but is it? Beware of the domino effect!

- It is estimated that as many as 50 per cent of all insects have been lost to the Earth community over recent decades. There are also anecdotal stories that indicate insects in specific locations are in decline, as evidenced by how many insects are killed on windscreens as people drive from place to place. It is a positive that people have at least noticed their decline, even if it is only because they have cleaner windscreens!
- With global warming and the changing climate, insects have nowhere to be safe. Insects evolved over many millions of years, so many insects are not able to adapt fast enough to new climatic conditions to survive. However, for some insects, they might adjust and flourish, to the detriment of crops because some of them are tough little creatures, like the cockroach. It is a survivor.
- Changes in crop management have threatened many insects above and below the ground through the use of chemicals used in farming practice.
- Climate change can affect food security for insects, especially if plant life is out of sync with insect life.
- The loss of pollinating insects through habitat loss or global warming is a great cause for concern for farmers, especially orchardists.
- Many birds and other creatures depend on insect prey for their food security.
- There are some pretty serious 'feedback loops' and 'domino effects' when it comes to insects. For example, no insects equals no pollination of crops and no food – not good for man or beast.
- The absence of research on environmental changes for insects means that long-term studies regarding the vulnerability of insects is not generally available. This is a serious omission given that Rachel Carson, the mother of environmental activism, blew the whistle on the effects of chemicals such as DDT on insects and birds in

her compelling and best-selling book titled *Silent Spring* (1962). While her work resulted in DDT being outlawed as a pesticide, other chemicals have been developed that have been equally deadly to insects.
- The loss of insects to the world would be catastrophic for every creature, including us 'big bugs'.

In summary, insect ecology is a concern for scientists studying them as insects play a vital role in the health of ecosystems, the pollination of plants, and cleaning soil and forest debris, and their service roles are well documented. The volatility of global warming and climate change is a real threat to their continued existence, along with habitat loss. Because of the diversity of insects, it is hard to pinpoint what is going on with them as a result of global warming because they are being impacted on all around the world. There appear to be many studies in progress on insects that give references for how many insects are declining in number and the movement of insects because of the changing seasons. For example, monarch butterflies, witnessed in their millions as they migrate, are thought to be in trouble. What seems to be agreed on is that global warming will wipe out many varieties of insects.

The loss of insects – and many are already lost – could bring about far-reaching consequences for biodiversity and therefore life on Earth. Insects are small and apparently fairly insignificant in the life of most of us, but they actually hold the balance and harmony of biodiversity in place. We, the human species, have been around for about two hundred thousand years compared to their millions of years; do we want to be responsible for their extinction? I talk to my visiting bees because they are pollinators, and I know they are in danger. Do we really want artificial food on our plate?

Bird Biodiversity

Ten thousand different species of birds fill Earth with dance and song, and they also give Earth amazing colour and beauty. Birds evolved out of dinosaur times, about 150 million years ago in the Jurassic Era, following the early flyers, the insects, into the sky. Birds inspired flight. People longed to soar through the skies with sheer delight, and as a result, all manner of flying machines have been invented.

Birds are creatures with feathers, which defines them as a species. The variety of birds is amazing, from very large, such as emu and ostrich, to very tiny wrens and finches. Some birds have forfeited flight for fleetness of foot and used their wings to turbo-charge their running over the ground, such as quail or, in the case of penguins, their swimming in the ocean.

Bird ecology refers to birds, plants, and animals as well as the environment that they have adapted to for their habitat and food security. The balance and harmony of creatures are reflected by how they coexist and their interdependence on one another. Charles Darwin, through many years of research, is accredited with the theory of evolution. Birds were one of his most studied creatures, and he discovered that special features in birds, such as the length and shape of their beaks, evolved to take advantage of food sources – beaks for sucking nectar from plants, beaks for cracking nuts, or beaks for catching prey – so they have evolved to be herbivorous or carnivorous for food security.

Many birds migrate each year, and some fly thousands of kilometres to a traditional food source. However, because the climate is changing, some birds find that their traditional food source has gone or has not arrived when they arrive at their traditional feeding grounds, and so they starve. Also, toxic food can impact on the quality of their eggshells, so chicks do not survive. Ocean birds have been known to feed their chicks pieces of plastic because they think it is food. Plastic cannot be digested by any creature, so this is now a real problem, especially for water birds that dive on what they see as food for themselves and their young.

Some birds migrate for seasonal climate but mostly for the availability of food, even though their migration paths can be perilous for them. Animals and other birds know of the migrations of other creatures, and so they wait to prey on them as they arrive for their food supply as part of the web of life. People and feral animals, even domestic cats, are also predators of birds, and so no bird is really safe. The evolution of bird life has taken millions of years, and the variety of birds in the world is astonishing. However, because of human activity, many species of birds are going extinct. So once again, the loss of habitats and food security because of global warming and environmental degradation are the main causes. People have encroached on their native/traditional territory, and some birds cannot adapt fast enough to alternative environments, or they may not be welcome in any other bird's territory. Like every other creature that is threatened by human interference, birds are also vulnerable, and bird species loss is being carefully researched.

What, in a nutshell, is happening to bird ecology?

- Birds have a role to play in cleaning up waste, such as eating dead animals, aiding pollination, controlling insect populations, and seed distribution as well as adding to the rich diversity of life.
- Many species of birds are threatened by habitat loss as land is cleared for crops or pasture.
- Bushfires intensified by rising temperatures also cause the destruction of birds' habitats and the loss of their fledgling young.

- Many birds depend on insects for their food security. If there is a loss of insects, then birds are threatened.
- Many species of birds are threatened by global warming and the changing climate.
- Migratory birds are the most threatened because they have adapted over time to arrive at their migratory destination where their food source will be waiting for them. However, changes in the climate can mean their food supply arrives too early or too late for them. This is already happening for some birds.
- Climate change means huge variables in weather that affect birds because birds are like all other creatures; they have evolved to live in certain temperatures. In extreme heatwaves, birds have been known to drop dead out of the sky!
- Humans are a threat to bird species because we have always killed them to eat or for their versatile feathers as well as their eggs.
- Birds are also threatened by other birds and animals, especially feral cats. If creatures like cats hunt them in the night, the birds have no defence. Research on cats shows that one million Australian birds are killed every day by feral cats, while domestic cats are also taking a toll on bird life, killing sixty million birds each year.
- Birds live and work within their ecosystems to help keep balance and harmony. There is a school of thought that says birds are vital to forests, wetlands, and grasslands. We simply do not reflect on what they do that is vital for life.
- One in eight birds globally is threatened with extinction, while many are already critically endangered.

We were born in a biodiversity-rich world. Everywhere, there are birds that astonish and delight us, and they are presented to us in all their glory on our television screens. They are the birds that have shared our evolutionary journey because they are the great adapters to the conditions of life. Any loss of biodiversity is a tragedy as every creature has an important role to play in keeping the biosphere in balance and harmony. A staggering number of birds – that is, 40 per cent – are in decline because of human activity, according to *'The State of the World's Birds: Taking the Pulse of the Planet'* by Birdlife International (2020). This document is a very sobering read and well worth a look because every one of us can do something to preserve birds in our backyard.

Animal Biodiversity

Animals fill our lives with sheer delight. In the wild, in our homes, or in a zoo, people love to share their lives with animals. Animals are the species we most identify with because our DNA is closest to animals. We share our evolutionary history and therefore have a great deal in common with animals. Our closest relatives are the chimpanzees, orangutans, and gorillas. Scientists tell us that human DNA is 98 per cent like the chimpanzee's. The 2 per cent difference is extraordinary.

The first animals lived in the ocean until changes in the atmosphere meant that they could venture safely onto land. No one really knows how many different species of animals there have been in the past and disappeared through various extinctions. Even today, we cannot name them all, but we know that different species of animals have been numbered in their millions. We domesticate animals and depend on them for our nourishment in the great circle of life. As well as being interrelated and interconnected with animals, we are also extremely dependent on them as they are a very important part of our food chain. For many people, animals also provide skins for clothing and shelter.

Humans are the dominant predator species, and no animal species is safe from humans. Animals have evolved and gone extinct over time as part of a natural process, but what is happening now is that all animals are at risk of extinction. When we allow our species to kill a rhinoceros just for its horn to make medicine or as a status symbol, we really must look at ourselves, shake our heads, beat our breasts, and ask ourselves, 'How are we allowing this to happen?' The fact that scientists have set up DNA banks to preserve the genetic material of endangered animals for the future is something to celebrate but something that should also alarm us. Why are they needing to collect and save the genetic material of animals? Are scientists so fearful for the future of animals that they are taking this precautionary measure of collecting and storing sperm, eggs, skin tissue, and blood samples in case of the extinction of some or all species?

There has been a great deal of research done on endangered iconic animals in the wild, such as polar bears, wolves, pandas, tigers, orangutans, gorillas, snow leopards, and elephants, just to name a few. Thousands of environmentalists are dedicating themselves to their preservation, but for some animals, human endeavours may not be enough to save them. Less known animals are also at risk of becoming endangered.

There are also domesticated animals, such as cattle, that are generating concern, not because they are threatened with extinction but because of the amount of methane they release into the atmosphere. Feedlots can be identified from space from their methane plumes. However, as human consciousness develops regarding our relationship with other creatures, ambassadors for animal rights have stepped up and become spokespeople for better conditions for animals raised in captivity for human consumption.

As well as the loss of habitats and traditional food sources, all animals are also at risk from the changing climate as they evolved to live in certain climatic niches that include the right temperatures. Some animals, such as goats, are already moving upward on mountains, but when they get to the top of the mountains, there is nowhere to go. What animals require is that we learn about them and understand that they have the same needs as us: a steady climate, a secure food supply, and a safe habitat to rear their young. Our world will be so much poorer if we lose any of our animal relatives.

What, in a nutshell, is happening to animal ecology?

- Global warming is impacting on animals in a visible way. The iconic polar bear is a classic example of the impact of climate change. For some polar bears, the sea ice from which they hunt is no longer there, or the ice is not thick enough to hunt from. In addition, their hunting season to prepare for hibernation is now much shorter to the point where polar bears are now going into villages to scavenge for food. So what happens to those polar bears? Do we spend thousands of dollars to relocate them hundreds of kilometres away, or do we just shoot them? How do the people who now have polar bears in their backyard deal with this issue? What can they do? No wonder scientists fear for the longevity of polar bears, which are on the seriously threatened list of animals. If Arctic ice disappears, as is predicted, then the iconic polar bear is doomed to extinction.
- Because of the changing climate and bushfires, another iconic Australian mammal, the koala, is potentially a threatened species. There are more bushfires because of more lightning because of global warming that is changing the climate. So the fear is that koalas may well disappear as they depend on special eucalypt tree leaves, and even eucalypt trees, Australian native trees, are dying because of heat or drought.

- If humans can suffer from heat stroke when exposed to extreme temperatures, causing headaches and nausea, what temperature can cattle in feedlots cope with before they succumb to heat stroke?
- Recent reports confirm that at least seven hundred animals are at risk of disappearing because of the consequences of climate change and habitat loss because adaptation for many is not an option.

While reading up about the plight of animals, one feels very melancholy. They are our closest kin, but the human species is decimating their habitats, changing their climate, and degrading the ecosystems they depend on. When park rangers are stunning elephants to remove their tusks to save the lives of the elephants from poachers, we need to be concerned. We are the educated generation of 'wise humans', so why do we feel the need to wantonly kill creatures in the wild for trophies or money? What is worse is that what we do to the other-than-human world, we will do to ourselves. No one wants to talk about the consequences of climate change on our habitats as well as food and water security, and they are not secure at all.

A good example of the vulnerability of animals or mammals is what happened here in Melbourne. Just before Christmas 2019, thousands of bats/flying foxes died when the temperature rose to 110 degrees Fahrenheit in Melbourne parks over a three-day period. It was a horror story, and it happened so unexpectedly and so quickly. In three days, when temperatures rose to beyond the flying foxes' ability to cope and keep cool, some 4,500 flying foxes perished in the heat. It is believed that at least 15 per cent of the flying fox colony died in that heatwave. This is an excellent example of animals that need a stable temperature to live, and we humans are changing the climate for them. We have air conditioners to help us in extreme heat, but if they were to fail because of a heavy drain on the energy grid, then perhaps our old and young will not survive extended extreme heat. Any mother of an infant knows how difficult it is to feed a baby so as to keep it hydrated in extreme heat. We just don't want to think about that scenario, do we? If you listen to any climate-educated person talking about the need to urgently cut global emissions, you can hear the information above in the voice of their hearts, not so much the words because climate scientists are reluctant to say it as it is. The same scientists are not doomsday speakers; they are the voice of reason for all the creatures that have no voice but theirs. We must believe in the 'power of one' to make a stand, and collectively, we can and must bring about a change in our relationship with the other-than-human natural world. No animal in the wild should go extinct on our watch if we want to preserve ourselves.

Reptile Biodiversity

Reptiles have been evolving for about 320 million years. It is thought that their ancestry belongs with the amphibian family. Some amphibians chose to stay largely as water creatures, while others opted to be land lovers, even desert dwellers. Having scales and producing hard-shelled eggs that they lay on land defines reptiles. The reptile family includes turtles, lizards, snakes, alligators, and crocodiles.

The crocodile is an iconic reptile that has adapted to live on land and in salt and fresh water. They prefer water. Crocodiles are survivors. During the course of evolution, creatures come and go, but the crocodile survived even the dinosaur extinction. It was only recently that scientists discovered the contribution reptiles make to the health of Earth. Every creature has a unique niche through interconnectedness. Every creature helps in some way, but humans may not yet know the vital role each performs for the health of the macro ecosystem. Human ignorance of reptiles and the role they play cannot be the measure for how we value them.

Scientific research demonstrates the decline in the number and variety of reptiles on a global scale, and the culprits again appear to be the changing climate, land clearing, and the subsequent loss of habitats as well as invasive species that the native reptiles have not had to deal with before. Added to these challenges are environmental pollution, disease, and, as for all other ecologies, the changing climate. The changing climate may be happening too fast for some reptile species to adapt to. Like other creatures, reptiles have adapted to climate and terrain over millions of years.

Reptiles have such a long evolutionary history, more history over time than we can possibly imagine, but today some are endangered because of thoughtless human activity. Indigenous people have always eaten reptiles as part of their diet, but this was done in a sustainable way with respect and gratitude. However, reptiles are being consumed in an unsustainable way as a commercial enterprise. Killing frogs for their legs, crocodiles for their skins, or turtles for their shells is the same as killing sharks for their fins, elephants for their tusks, or rhinoceros for their horns. This is completely unsustainable and a direct pathway to extinction for these creatures. Our awareness begins with education about threats to all biodiversity, and once people become aware of any endangered species, they tend to transfer their concern to other species as well.

What, in a nutshell, is happening to reptiles?

- Ecology scientists who are assessing the state of reptiles have concluded that 20 per cent are threatened with extinction, 10 per cent are deemed to be critically endangered, 30 per cent are endangered, and 40 per cent are believed to be vulnerable. The race is on now to

- ascertain the true plight of reptiles and what can be done to mitigate their demise.
- Reptiles are found pretty much everywhere except on the ice, even though they are cold-blooded creatures. They are extremely sensitive to the changing climate as their bodies are adjusted to specific temperatures within their environment.
- Scientists are still researching how reptiles such as lizards are affected by global warming. It seems that their survival rests on access to shade from trees or outcrops of rocks when temperatures rise above normal levels.
- Reptiles are extremely vulnerable to the loss of habitats through human exploitation, especially the rainforests. As the trees disappear, the consequence is that the reptiles that call rainforests their home will disappear with them.
- Many different reptiles have carved out a special niche for their safety and the security of their food supply. However, the changing climate can impact and is impacting on their traditional food supply. One major study has already shown the effects of climate change on at least half of amphibians and reptiles.
- The domino effect of losing insects will seriously influence the survival of some reptiles.
- Reptiles are creatures that can disappear without humans being aware until it is too late to save them; therefore, they are extremely vulnerable.

Reptiles are not part of our everyday life, so we are not, in general, concerned about how they are doing in the changing climate. What we do know is that reptiles have evolved with us and have a role to play in keeping the balance and harmony of the ecosystems supporting life. They are vital in keeping insects and rodents in check as part of the web of life. Likewise, they are prey for birds and other reptiles, and they help to clean up carrion, which helps prevent disease. Frogs are thought to be like 'canaries in the mine'. A healthy ecosystem will have frogs to fill the air with song at night while they demolish mosquitoes, thus saving people from malaria. Turtles are also at risk, not just from global warming and ocean warming but also from human exploitation and pollution. We have all seen photographs of turtles entangled in fishing materials, especially ghost nets. Turtles lay their eggs on beaches, so now they are in trouble because of rising oceans, and their evolution has programmed them to return to the beach where they hatched to lay their eggs. How do they choose another beach? Turtle survival statistics are not impressive at the best of times given that only one in a thousand will make it to adulthood. Even as eggs, they are extremely vulnerable to marauders that seek to devour them as the journey from hatching to the water is perilous for them.

Perhaps it is time to give some thought to the wonder of reptiles and amphibians before we lose them to extinction, whether it is a crocodile or a skink. Reptiles would be grateful if we could reprogram ourselves to value each creature for its own self and be a voice for all creatures that need our protection. We cannot let 370 million years of evolution disappear because we are heating the planet and not sufficiently concerned about the welfare of reptiles. Once enlightened, we have to care for reptiles as they have no voice but ours.

Ocean Fish Biodiversity

The ocean covers two-thirds of Earth's surface, so it is not surprising that it had more wildlife in it than the land. The ocean was once full of wildlife. The food chain for creatures of the ocean is delicately balanced, beginning with plankton as the baseline up to the top predator of the sea, the shark. The combined fish ecology has unimaginable varieties of colours and shapes, and photography shows us what we may never see as land dwellers.

The people who have been able to dive in the ocean have shown us an awesome wonderland, the paradise home of fish. The ocean and its wildlife are a common heritage for all people that have been, for many millions of years, creating their complex ecology.

Most fish live within the ecosystem of the ocean. Within the ocean, there are different habitats, such as the open sea, deep and shallow water, and coral reefs or kelp. There is a vital connection between fish and their biodiverse environment as their aquatic habitat provides safety and food security. Fish have a role to play in nutrient recycling and cleaning the ocean, and they are a good example of 'one creature's waste is another creature's wonder food'. The web of life in fish ecology is astounding and perfectly ordered through evolution.

Fish have provided food for many millions of people. For islander people, fish are their staple diet, providing much of the protein and omega-3 fatty acids needed for their health and well-being. Today the balance is shifting, and many fish are endangered, creating an imbalance in the food chain in the ocean. Before the Industrial Revolution, people could eat only what they could spear; then hand-thrown nets were used, followed by light boats with handlines or small nets, followed by larger boats with longer, stronger lines, followed by fishing trawlers with bigger and better nets. Today there are factory ships with massive nets and sonar and plane spotters that capture everything in their path. No fish or aquatic creature is safe in the ocean today. Humans are now the top predators of the ocean; we have superseded the shark. Commercial fishermen are extremely efficient and insatiable hunters as fish are an extremely valuable market commodity, and financial greed is now the prime motivator for fishing.

Oceans and their inhabitants are now a threatened species. The fish populations have gone from infinite to finite, renewable to non-renewable, sustainable to non-sustainable, and inexhaustible to exhaustible in fifty years. The North Atlantic cod, one of the most amazing fish of the sea, was fished to near extinction until a moratorium was placed on them in the 1990s, but it appears they are not coming back since their population was virtually decimated. Scientists call this an ecological catastrophe. The bluefin tuna, found around the world, is now in a delicate balance, with a decline of 80 per cent of its population. The problem is that after the bluefin tuna, the next biggest fish will be targeted until all big fish are gone. Beginning in the 1950s, fishing became big business, and now aquatic scientists maintain that as much as 90 per cent of big fish are gone. We will have eaten them or wasted them as by-catch.

What, in a nutshell, is happening to ocean fish ecology?

- Fish are an endangered species because of the world's biggest predator, humans.
- Global warming has had a negative effect on fish reproduction. Fish stocks have been affected worldwide.
- Overfishing has depleted some fish species, thus changing the food chain in the ocean.
- Warming river water is influencing salmon spawning behaviour. The outlook for salmon is not good, especially if there is a heatwave as they swim to their spawning grounds up the rivers. Humans had already put obstacles in their way, like dams.
- Scientists say that humans are at war with fish, and humans are winning. They predict that species after species will collapse, changing the food chain in the ocean. Marine scientists predicted the demise of the North Atlantic cod, and that prediction is now realised in our time.
- One-third of fish species are in the state of collapse.
- Although there are fish catch limits that are regulated in most countries, at least 50 per cent of fish taken from the ocean are considered to be illegal.

- The by-catch that nets collect are thrown back into the ocean dead. At least 10 per cent of a given catch is discarded and therefore wasted. This is considered to be a crime against nature. Killing reef sharks just for their fins and throwing them back into the water alive to die is also a crime against nature.
- Farming fish is not the answer as it can take five kilos of wild fish to make one kilo of farmed fish. Where is the wisdom in that little equation? Who are we kidding?
- With the loss of one species, there is a knock-on effect. As sharks are an endangered species, rays have moved up the food chain and are far more abundant, so now they are being captured as a food source for humans.
- The overfishing of krill in the Southern Ocean provides bogus vitamin supplements to cure every kind of conceivable ailment for the human species while depleting krill stocks that are vital for whales and penguins that depend on them.
- Only about 1 per cent to 3 per cent of the ocean has marine reserves set aside for fish safety, but since fish are wild, they wander outside their reserve. Marine scientists agree that at least 30 per cent of the ocean needs to be set aside to rejuvenate fish populations before it is too late.
- The international goal to establish marine reserves was 10 per cent by 2020. Are we on track, I wonder?

Where I live, I cannot see this devastation of ocean wildlife. When I walk on the beach, I cannot see potential fish ecology collapse. As I walk through my local fish market, where there appears to be a bountiful variety of fish, it is difficult to get it through my head that all of the above is happening, but when you learn about the tonnage of fish caught every year, you have to believe that this cannot go on ad infinitum. The insatiable need for fish for this generation of people is going to tip the balance of wildlife in the ocean, which will be catastrophic for marine life. Fish populations are decreasing, and the prediction is that there will be total collapse in the not-so-distant future – that is, the total collapse of an ecosystem that will have been squandered into extinction.

At every stage of the fish food chain, aquatic creatures are being depleted, from krill to sharks. Whales have always been fished, but there are determined environmental groups, such as Greenpeace, that have been heroic in their efforts to stem the tide of whale extinction. If we could think of the big fish of the ocean as lions, elephants, and tigers on land that we are prepared to defend, then can we not defend these mighty, iconic creatures of the sea? In Australia, when the discussion of culling invasive species such as camels, brumbies, pigs, goats, and buffalo comes up, there is usually an outcry from the populace, probably because we actually see photos of these animals and how they are slaughtered, but we don't see ocean fish being captured and killed. It appears to be a case of 'What the eye does not see, the heart does not miss'.

There is no negotiating with Mother Nature when the harmony and balance of nature are disturbed, so it is likely that there will be more extinction of fish species. Even the heads of the ocean food chain, the lions of the sea, the sharks, are in danger. It is quite possible that more swimmers and surfers will be attacked by sharks as they come closer to the shores to feed because we have eaten their food. After every shark attack, in the past, humans hunted down the culprit and killed it along with any other shark in the vicinity. We have upset the balance of the food chain in the ocean, and therefore, there will be consequences.

In summary, we are bringing about the sixth mass extinction of biodiversity that has shared our evolutionary journey. All of the above biodiversity has registered extinctions of some of their species, even the rivers. The statistics are alarming and deeply regrettable because there is no coming back from extinction. Human activity is so invasive of the natural world, and it is horrifying to know that no creature is safe from the main perpetrator of violence against them – us. Much of the environmental discourse is about global warming and climate change, but there is nowhere near enough discussion on the plight of creatures in the natural world. We are facing the sixth mass extinction of life on this planet, and we will be responsible for it. It is happening at rapid speed now on our watch.

We have to ask ourselves a question: do we want to be the species to sign the death certificates of so many creatures that have journeyed with us through evolution given that no creature is safe from us, the most prolific and dominant species on Earth? We must ask ourselves – *Homo sapiens*, wise humans – whether we really are wise or just stupid and extremely thoughtless. The plain fact is that we are dependent on biodiversity, even though the other-than-human biodiversity would do perfectly fine without us. We are the needy creatures. The feedback loop to human well-being is that we cannot survive as a species unless the rest of the natural world survives and flourishes with us.

The loss of biodiversity will have a domino effect, and we may well be the last domino to fall. I realise that sounds extremely dramatic, but it's the reality we must face. Remember – we are members of the animal kingdom, just big bugs on the earth, and absolutely dependent on the natural world for our survival. Finally, ecological scientists are furiously clanging alarm bells, and we must listen and respond. Our lives will be so much poorer if we lose just one more species from Earth's community of life because of our thoughtlessness. We do not think about biodiversity; we fail to understand and appreciate the biodiversity that enriches and supports our lives. Is it possible to change our dominant relationship with the natural world before it is too late for the recovery of threatened biodiversity?

CHAPTER SEVEN

The Laborious Turning of the Anthropocene Era

As we contemplate and seriously reflect on what has brought us to an environmental emergency, we can feel disempowered because without thinking, we have bought into the idea that our progress towards a good life depends only on our own endeavour – that is, if we work hard enough, we could have it all. Now we must turn that thinking around, but we cannot do it alone as individuals. Just as it has taken time to get to this stage of awakening to the needs of the natural world, so will it take time to reverse the destruction we have caused. Each generation can continue to reflect on and review their progress towards restoring the Earth. If the whole of humanity comes together through education to help us become environmentally literate then we are in with a chance not only to renew the face of the Earth but also to enable ecosystems and biodiversity to continue their progress towards an evolutionary wonderland for all creatures, including us.

Anthropocene Era

Throughout history, different names have been given to different geological periods in the history of Earth. The Holocene Era of the history of Earth takes in the last twelve thousand years since the last major Ice Age. Throughout this period, the global temperature remained stable and has only deviated by about one degree. We are leaving the Holocene Era in favour of the Anthropocene Epoch, created by human domination of the Earth.

Anthropology is the scientific study of the human species. Anthropocentrism means it is all about us. We are anthropocentric in that we do not, in general, think of the needs of the rest of the natural world; we are mentally neurotic in separating ourselves from the rest of the biosphere, so we can do what we like to all other creatures. The ultimate end of this separation is the destruction of the natural world, to the point where we may find ourselves in a feedlot fed with artificial food. I know that sounds harsh, but if we continue to destroy or devour biodiversity at the current rate, what is the alternative?

From a geological point of view, the Anthropocene Epoch is one of changing the climate and the geological formation of Earth through human activity. This term is not yet part of any scientific classification, but surely, it is now possible to discuss it because it has been/is a period of incredible change to the Earth. Basically, the Anthropocene Epoch is the period that covers human domination of Earth. It is recent because it began with the Agricultural Revolution, approximately ten thousand years ago, with domination of the land. It continued through the Industrial Revolution, which began just over two hundred years ago and required massive amounts of raw materials to continue churning out more consumables for our society to dispose of in landfills. The Anthropocene Epoch is now evidenced by complete domination of Earth through the Technological Revolution, beginning over sixty years ago. The Anthropocene Epoch has taken domination of nature to a whole new level because of advances in machinery and other inventions that collect Earth's 'resources' at an insatiable, unprecedented, and quite clearly unsustainable rate.

The Anthropocene Epoch is one of human design because we continue digging, blasting, sucking, chopping, polluting, spewing, fishing, damming, hunting, bombing, clearing, burning, draining, bleaching, and whatever other activity demonstrates human domination of Earth's ecosystems. Every gain for economic progress comes at a great loss for Earth, and because many of Earth's gifts are finite, they are irreplaceable. Human domination impacting on ecosystems and biodiversity is well documented, and there has been a great deal of interest directed towards threatened, endangered, and extinction of species because of human activity. There is a consensus amongst ecologists that around 50 per cent of biodiversity has been lost to the Earth community in the last forty years. The animals and birds that are still flourishing are well away from the human environment, so they are safe for the time being. However, global warming will still take a toll on them. The devastating loss of biodiversity is to our eternal shame as these are the creatures we know; they evolved over millions of years with us. Many of the creatures that are being lost are so much like us in their DNA; thanks to advancement in science, we understand this important feature of our interrelatedness. We are the first humans to know this; we understand the evolution of species. Does that mean anything to us? Other creatures would see us as just big bugs on the Earth, nothing special to them, just an inconvenient nuisance that they can't get rid of because their lives would be so much better without us. The only way to improve our relationship with the natural world is to power up the love for our home planet, and we can do this!

The changing climate is another cause for concern. Scientists tell us that anthropogenic climate change is brought about by increased carbon dioxide and other toxic gases in the atmosphere through human activity. They have been able to measure changes to the amount of carbon dioxide and toxic pollutants in the atmosphere since the beginning of the industrial era. Scientists who study environmental changes in Antarctica have drawn out ice core samples that record a timeline of pollutants in the atmosphere, so there is no mistake about the effects of the Anthropocene Epoch, no faith required; it is all self-evident. The ice does not lie.

The Anthropocene Epoch has allowed the gifts of Earth to be used and abused without consideration for the general well-being of the Earth itself and future generations. When it comes to justice for the whole community of beings, it is the Earth itself that requires ethical environmental justice. We humans have largely, indiscriminately trashed the biosphere for our own purposes over the last two hundred years. Ecosystems and biodiversity do not have a voice to inform us about the changes they are experiencing because of our needs and wants. If the Earth and biodiversity could speak to us, then they would share some ecological wisdom with us, such as 'We do not need the human species. The human species needs us, and this is how you are treating us'.

Planet Earth and its community of non-human beings are currently

- violated,
- humiliated,
- unhealthy,
- marginalised,
- powerless,
- exploited,
- voiceless,
- dominated,
- suffering,
- degraded,
- subjugated,
- oppressed,
- demoralised,
- misused,
- manipulated,

- wasted,
- dishonoured,
- despoiled, and
- threatened.

Each of the above descriptors can be expounded on at length. The degradation of Earth is not something we consider daily, so for our own health and well-being, we could consider giving the Earth a hug today by embracing one of its wonders. It is difficult to think about what we, as the human species, have done to our awesome home planet, its ecosystems, and its biodiversity over the last number of decades; however, a new era is dawning. The Environmental Revolution is underway, and people everywhere, numbering many millions, are becoming really Earth literate and Earth conscious. Many people have undergone their own ecological conversion, and there is no turning back. Once you really understand our relationship with the total Earth community, then you are committed to preserving that which you love. Change is in the air. For some, it is called the Ecological Period, when we are well informed and deeply committed to make a stand for the health of our home planet, ourselves, and all of its biodiversity in perpetuity. It is a new era of awakening to the Earth.

Ecological Period

The Ecological Period is an awakening to the history of Earth and the evolution of creation. *Eco* means 'home', so it is the study of our home planet. The Cenozoic Era of the last sixty-five million years is giving way, although reluctantly, to the Ecological Period. The Cenozoic Era was a period of explosive biological development and evolution that produced a most extraordinary diversity of creatures, including us that make up the totality of Earth's community of life. 'Ecological Period' is a term given to a momentous change in understanding our relationship with Earth.

It is reclaiming what we have lost as our lifestyles, especially in cities, have become severely disconnected from the natural world. Over 50 per cent of people live in cities, in the built environment, largely removed from the natural world. Until children learn otherwise, milk comes from bottles and eggs from cartoons in their supermarkets.

The Ecological Period is founded on three major principles of environmental consciousness, which include the acceptance of humans' interrelatedness with, interconnectedness with, and absolute interdependence on the natural world. Everything is part of the great Earth community of beings. If most people in First World countries could grasp that fact, then we are in with a chance of saving ourselves. For this reason, the Ecological Period will most likely be the most profound and critical era in human history. This is the time for the human species to embrace the reality that Earth is an interlinked community that includes us and all other creatures. Every creature is integral to the Earth community and important to the health of Earth's biodiversity and biosphere. The folly is that the human species sees itself on top of the pinnacle of creation and all other creatures as subservient to us. The truth is we are all part of the great circle of life as a community of all beings. However amazingly superior we think we are, we cannot live without the rest of the natural world, and we are destroying it at an unprecedented rate.

The Ecological Period notes a turning point from the thoughtless human domination of Earth's resources to respect for Earth's gifts. For some time now, the human species has been acting as though it is in charge, but Earth is showing signs that our activities and total lack of respect for ecosystems supporting life is affecting the well-being of the natural world. It is a time for listening to Earth. Mother Nature, through the Earth community, is constantly showing us how we can be in the right relationship with our home planet. There is only one Earth; everything we need to live is on this planet. There are enough gifts of Earth for all, but we must rethink how we use and reuse the gifts of air, water, soil, ocean, flora, and fauna forever. Humans make a huge mistake by thinking that they are more important than the natural world that supports them. Have you ever tried living without air or water?

One way we can all enter and embrace the Ecological Period is to celebrate the mystery and miracle of life on Earth. We who are educated realise that we live on an astoundingly breathtaking but fragile planet. Mathematics teaches us that Earth is not limitless in its ability to provide for every creature if there is waste and the degradation of vital earth systems, including its wildlife biodiversity. The profound realities supporting the need to move quickly into the Ecological Period are as follows:

- The Earth community is interrelated because every creature shares a common ancestry. DNA discoveries demonstrate that the common house mouse is 97.5 per cent like us; 2.5 per cent turns mice into humans! Really, that does make a mouse scary! All creatures need the same healthy ecosystems as the human species.
- The Earth community is interconnected because every creature shares the same soil, air, and water. We are drinking the recycled urine of a dinosaur and breathing the same air as woolly mammoths. The soil provides a habitat for many micro-organisms and produces food for a multiplicity of creatures, including humans.
- The Earth community is interdependent on every ecosystem or biome. Mountains and valleys, rivers and creeks, ice and snow, ocean or land – everything is interlinked. Nothing on Earth functions independently or acts alone.

The Ecological Period recognises that Earth is a community of beings and that every being is vital to the health of the whole community. Just because we do not understand what each creature contributes does not mean that they are not playing their role to keep Earth healthy. Before starting to write today, I looked at a tweet on a sea cucumber. The heading for the two-minute video was *'Sea cucumber poop is surprisingly important to ecosystems'*. Well, now it seems that this little ocean creature, which does a huge poop, is very busy picking up sediments on the ocean floor and cleaning it; its poop fertilises sea grass and helps to prevent algae blooms that deplete oxygen in the ocean, and amongst other things, it helps with the acidification of the ocean. I think this little creature does more for the health of the planet in one poop than I can do in a day or ever!

We know-it-all humans are, are in general, ignorant about the skills and contributions every creature brings to keeping Earth healthy. I have studied the dung beetle, for example; now that is an awesome insect that does a great job cleaning up after other creatures, so a big thank-you goes out to the dung beetle. We are living in a time of awakening to the needs of the total Earth community. We can and must wake up to exactly what ecosystems and their biodiversity of living creatures do to keep our home planet healthy; perhaps then we might feel empowered to protect all creatures. Humans are just one species within the Earth community of beings, but some of us are very needy and very greedy creatures. Our ecological footprint clearly demonstrates that fact, and we will examine our ecological footprint a little later in this narrative.

CHAPTER EIGHT

What Are We Thinking?

Business as Usual

To address the issues above, we need those who have the power to exercise it always in favour of the health of the Earth and its biodiversity. However, with slogans dominating the airways like 'jobs and growth', 'Build back better', and 'Economic growth is the steering philosophy', the Earth will have little chance to renew itself. Many countries have a growth obsession dominating their mantras for a bigger, better future, but unlimited growth will not be a panacea for anyone. In fact, the philosophy is downright dangerous for all creatures and potentially destructive for life-serving ecosystems. We are the guests of Earth, not its masters.

'Business as usual' has developed its own methodological paradigm that is dominating how we live on Earth. It has its own vocabulary, such as 'economy', 'bottom line', 'free markets', 'currency', 'productivity', 'resources', 'enterprise', 'growth', 'debt', 'recession', 'capital', 'progress', 'GDP', and 'globalisation'. We have a boom-and-bust mentality that when all of the above is in full swing, we are booming, but when the above is not working, we have the bust. All efforts are put into keeping the booming humming along, fired by an insatiable need for raw materials for jobs and growth, and this is all that seems to matter to those who created and perpetuate the 'business as usual' construct. There is no reckoning that the construct ledger is skewed towards environmental bankruptcy. We are drawing on the natural capital of Earth as never before in the history of the human species. Nature's bottom line is that perpetual progress along the 'business as usual' construct mentality is an illusion and that the propaganda machine has hypnotised us into believing that unlimited growth is possible. Mother Nature would say this deluded mentality is suicidal as there is only one Earth, one lot of resources/gifts to support every creature in the biosphere. Those resources/gifts are, for the most part, limited and finite. Not only are they limited and finite, but also, they are being abused and squandered as though there is no tomorrow to be concerned about. Who is overseeing nature's balance sheet? How close are we to being in the red?

Let us engage in a conspiracy theory. It is the 'myth of progress'. It is the myth of economy versus ecology and that the economy should always win. The myth is that the human species can continue to take from Earth everything it wants and needs without bringing about its own doom. The 'business as usual' paradigm is flawed because Earth is a closed system. We have whatever there is in the Earth bank but nothing more. It is a most salutary thought that my generation, over the last fifty years, has used up more of Earth's gifts than all previous generations combined. Let that revelation sink in if you want a motivator to be proactive for the health of Earth. It is called by Dr Joanna Macy, a well-known environmental educator, the 'great unravelling' of the fabric of life. The Ecological Period must be about the 'great turning' from the destruction of Earth's ecosystems and biodiversity to the restoration of such life-sustaining ecosystems and biodiversity.

The reality is that the powers that be, believe the human construct of 'business as usual' must prevail at any cost, and the cost is not being reckoned with, but we can change that because it is a human construct. Our economy is out of sync with ecology. We cannot go on bullying nature to give us everything we want without recognising the fact that the other-than-human biosphere has needs of its own. The world of nature does not have borders or boundaries, it does not have governments, and it does not give speeches or tweets; it just does its thing to keep the balance and harmony for the lives of the creatures that have evolved to live in it.

Countries and governments brag about their exponential economic growth, which comes at exponential Earth loss. Economic growth requires continued consumption, and we cannot go on consuming the gifts of Earth. We must find another way, and that must involve recycling everything. The basic flaw of the 'myth of progress' is that an extricating economy is a life-threatening economy. When forests are cut down, for example, the conditions that grew the trees in the first place may no longer exist because of the changing climate. There is a huge imbalance when economic prosperity is weighed against environmental bankruptcy. There can be no counting profit without balancing it against what is lost to the Earth.

Every advancement in technology to feed our consumerist mentality seemed like a good idea at the time. For example, globalisation seemed like a good idea at the time – to send goods here, there, and everywhere, only to return the goods to their place of origin in a can. Is this not madness? All this travel comes at a cost to Earth's air pollution. Another more basic example of humans' lack of sensitivity to the natural world is that some councils have put in doggy bags for dog walkers to collect their dog's poo when they walk on the beach. *Great idea*, you might think, but if the owners of the dogs throw the plastic bag of poo into the ocean to dispose of it, which I have witnessed, that is a double whammy for ocean life. Keep the beach clean for humans, but don't worry about life in the ocean! Again, 'away' is not 'away'; it is somewhere and our 'away' is often dangerous for other creatures. A message came to me the other day that we need to cut the strings on our Covid masks before disposing of them because birds have been found with the masks attached to them!

It is a strange paradox that the world has achieved so much that is amazing and good but without a collective conscience as to its impact on Earth's biosystems. We are giants of innovation and creativity but infants when it comes to Earth ethics. A new ethics paradigm is required, one that reshapes our place in the world within the confines of ecological needs and no-go zones in nature. If the natural world is only valued for its economic benefit, then we are in deep trouble. The 'use and abuse' mentality to nature's gifts will reap its own outcome, and we will be the losers.

Nature does not just have an economic value; it also has a spiritual value, an aesthetic value, an ecological value. If only there were schools that taught children about the complexity of nature and its importance to our health and well-being other than the goods and services that nature provides! Is it possible to bring into our understanding of Earth ethics a love for our home planet and a sensitivity to its needs as well as be humble in acknowledging our utter, absolute dependence on the gifts of the natural world?

A 'business as usual' mentality requires serious examination to create a balance that is sustainable – that is, human needs carefully measured against Earth needs. To do this, we have to bring about a cultural consciousness, a cultural revolution that is awakened to our deepest desire – that is, our survival as a species and the survival of all other creatures that have shared our evolutionary journey. We know what bees do to support biodiversity, but can we say with conviction what exactly the human species does to support biodiversity? Are we found wanting in this challenge?

Are we just serial consumers of Earth's gifts? We need a huge paradigm shift to a new understanding of 'progress', especially one to do with our waste! Nature has its own methods of taking care of waste – that is, one creature's waste is another creature's necessity for life. There is no hierarchy in nature; every creature serves the good of another.

So where are the changes that we need to make a difference to our ethical relationship with the natural world? Fortunately, there is an ethical environmental paradigm that can provide a pathway for discernment from what we have already learned.

- We can engage with teachings from indigenous people who hold the view that we cannot own the land. We are not masters of Earth. We are guests of Earth. For the time we are here, we are earthlings, we are closely related to the rest of the natural world, and we are interconnected with and completely dependent on the other-than-human world. We are wise humans; therefore, we are called to be faithful custodians of the land and sea.
- We are called to an ecological conversion and accept the challenge to live within Earth's ability to take care of us. We must downsize our ecological footprint.
- As custodians of the land and sea, we must be prepared to defend the whole biosphere from human domination.
- Environmental scientists have informed us that our present lifestyles are unsustainable, and so they have pleaded with us to inform as many people as we can about the health of Earth. To do this, we must continually educate our ecological selves.
- We are called to believe the science of climate change and then become proactive in bringing about the changes necessary for the survival of all life. To do this collectively, it may require joining one of the million environmental action groups.

- We must educate our communities on the need to recycle everything so that there is no such thing as waste. Recycling our waste will limit the need for raw materials.
- We must move to renewable energy as quickly as humanely possible to stop greenhouse gas emissions into the atmosphere.
- Every decision from governments needs to be made with the health of Earth in mind, not just for the good of humanity.
- All farmers must engage in regenerative farming to preserve grasslands and soil.
- We must challenge food outlets to divest themselves of anything plastic and be prepared to market food that, to the human eye, is not perfect; for example, a crooked carrot is just as healthy as a straight one.
- We must be prepared to challenge the status quo when it comes to government inaction to defend ecosystems and biodiversity as well as call out any action that does not consider the health of Earth first.
- We must nourish our own Earth spirit to engage in deep ecology with all creation.

There is a misconception that if we act on protecting and restoring the ecosystems supporting life, we will be worse off; this is not so. On the contrary, if we persist in destroying life on the planet, then we will be responsible for turning it into a wasteland. If we address the health of ecosystems and biomes, then Earth can rejuvenate into a wonderland to provide everything the whole Earth community needs. We don't just need a healthy Earth for the necessities of life; we need it for our collective souls, our collective spirits. For example, patients recover best in hospital if they have a natural vista outside their window.

Psychology of Denial

When I was growing up, it was accepted that we did not speak about religion or politics if we wanted to hang on to our friends. A newbie can be added – 'Don't talk about climate change!' People, in general, are nervous regarding talking about anthropogenic climate change. Both sides hold very strong views, and neither side, in general, seems to be convinced otherwise by logical persuasion. A new social science is currently being developed to explain why the subject of climate change is cause for such strong emotive language and divisive arguments.

Ecological anxiety comes from a psychological response to education about environmental issues. Some feel guilty that the devastation inflicted on the environment has come on our watch and are fearful of the future. Many feel helpless to do anything to turn the situation around. As well, there are still many who are in denial that it is as bad as it is. However, there is now a societal awakening, not so much because of the science but probably more about our young people making a stand for environmental justice. Young people are leading an outpouring of eco-anxiety over their possible future, and they are making their voices heard across the world and across governments. They are demanding intergenerational justice because they understand their rights to inherit a habitable planet. The shock jocks of denial are now threatened by these young people making a stand to the point where they feel they must ridicule them, but the young people will not be stopped. They know their lives depend on securing their future before 2030.

The question for this topic is 'Why, when the evidence is so strong and so many agree that this is our greatest problem and therefore greatest challenge, are we doing so little to stem the tide of global heating and environmental degradation?' This must be even more frustrating for climate scientists who have experienced a worldwide effort to curb the spread of COVID-19. Over the last number of months, billions of dollars have been put in place to support people financially, many millions of dollars into medical hospitals, testing, and inventions of medical equipment and vaccines. Rules, regulations, shutdowns, lockdowns, isolations, quarantine facilities, fines, and state border controls and closures that are unprecedented in living memory have been spontaneously put in place to stem the tide of the spread of COVID-19 infections and deaths, yet action on the changing climate has constantly been delayed because we can't financially afford to move too quickly, even given that we have known about this societal dilemma for over forty years! Sadly, the consequences of human-induced global warming are still going to take us by surprise.

To add insult to injury, the constant refrain from government officials regarding COVID-19 is that we must listen to the medical scientists; we must listen to their expertise and advice. Fortunately, in Australia, we listened to the scientific medical advice and vaccinations should protect our future from devastation from future zoonotic diseases. I can imagine that many climate scientists are wringing their hands in frustration as they do not get why there is so little action on climate science, why governments are not listening to them or why people, in their millions, are not up in arms protesting. Not listening to climate scientists and acting on their expert advice puts not only humans at risk of illness and death but also all biodiversity. It is important to note from stories of big social changes in history, such as women's right to vote or the abolition of slavery, that there is an ethical and strategic process for change.

- Any big change in history begins with denial that there is a problem; in this case, it is denial that we are in an environmental crisis/emergency.
- The change being presented to the community is then subjected to a barrage of ridicule directed north, south,

east, and west of the issue and everyone proposing climate justice action. Scientists calling for action have been effectively silenced as the 'fake news' people seem to have more access to the airways.
- Then there is a rise in violent opposition to addressing environmental issues plus those proposing action, and we have witnessed a monumental number of fake reasons for inaction; for example, we cannot financially afford to address climate change. The balance sheet for inaction will be exponentially greater and in the red.
- On the matter of addressing global heating – that is, the changing climate – there will eventually be self-evident acceptance that we are in an environmental crisis as environmental disasters become more frequent, more severe, more life-threatening, more destructive of the human-built environment, and more damaging to the economy as each environmental disaster is addressed. Thankfully, there are many millions who have reached this level of acceptance of an environmental emergency tending towards an environmental catastrophe.

Most climate scientists, environmental educators, and environmental activists know this process well and have experienced a barrage of ridicule that is so strong, it has silenced many climate scientists. They do not want to receive death threats to themselves and their families for their truth telling. The greater tragedy is that they are fearful of their own messages; they know that what they have to say from their research modelling and observations is a threat to those who will hear it. No wonder they are reluctant to tell it as it is! Climate change campaigner and former vice president Al Gore knows the above very well; has there ever been a man so maligned for so long who just keeps on keeping on with his climate message, knowing that it is an inconvenient truth? The same can be said for Leonardo DiCaprio, famous film star, but more importantly, he has been a champion of action to address climate change since he was a young man. He is also an advocate for saving creatures in the wild through the work of his foundation. Both men, well known to the public and admired by many, have been greatly criticised by media that is funded by big coal, oil, and gas. These shock jocks cast doubt on their motives and their expertise, which generates doubt in people's minds sufficient to stave off action for addressing climate change. Although they have been ridiculed, humiliated, and made a joke of by some deniers in the media, both men soldier on because they have seen with their own eyes the devastation of the planet. They have got off their safe decks and pursued the truth to validate their urgent environmental messages.

Recently, I watched a presentation by Dr David Suzuki, who is now in his seventies, titled *Unstoppable: Why It Is Time to Think About Human Extinction*. He has been a relentless campaigner on action for climate change for over forty years. He is a grandfather now, and as he says, he has nothing to lose, nothing to be afraid of, and no job to lose, so he chooses to be truthful and outspoken about climate science. One important statistic to comment on is that he believes we only have a 5 per cent chance of turning global warming around if we do not seriously address it now. He is in very good company, but sadly, his message makes hard hearing to those who are vehemently in denial of human-induced climate change. So how do you talk to those who are in denial of human-induced climate change?

- Be kind and understanding because it seems we all come to this discussion from extremely emotive positions.
- Try to find something environmental to talk about that you can agree on.
- Hold to your own view to provide food for thought, even if your position is not immediately acceptable.
- Be sensitive to the truth that we are all subject to the consequences of the changing climate.

- Ask them to share with you their story of how they came to their position on climate change.
- Do not pontificate on the worst climate science scenarios. Allow the media to perform that task.
- Agree very gently to disagree as it does not matter what you think or what they think; climate change is now indisputable, and it can speak for itself.

As someone who has been maligned for talking up the need for action on the changing climate, my approach now is to be very conciliatory and say, 'I hope you are right and I am wrong because I like your climate future scenario much better than my own. I hope with all my heart that you are correct in your judgement, that my grandchildren will enjoy the fullness of life, and that it is all a terrible conspiracy inflicted on ordinary people who are living in a state of confusion.' I also confess I no longer bring climate change up in certain company because I like the people I am usually talking to very much; however, there are quite definitely stages of denial. We can expect widespread denial when the enormity and nature of the problem, which is unprecedented, becomes obvious and people have not developed cultural mechanisms for accepting the reality of climate chaos. In Australia especially, there is an entrenched understanding that the climate and seasons are cyclical and variable, usually depending on El Niño; this is true, but there are other scenarios that must be considered, like the global rising temperature! People just do not join the dots that it is not El Niño per se that is responsible for changing the climate; it is the continual heating of the planet, which is measurable.

I know people whom I love and respect who are in denial that we are in a climate emergency. They think climate scientists are not worth listening to, that they are all about promoting themselves and their jobs by drumming up the need for grants. I also have friends in the environmental movement who simply do not believe we have a future as a species. They are of the firm conviction that we have already set up our own downfall, that there is no hope for us, yet they go on struggling to make the plight of Earth known and call on everyone to rally around to restore balance and harmony on the planet. Whatever view we hold, we must cling to the fact that we are intelligent people, that we are, in general, good people who want the very best for everyone.

Ecopsychology helps people who are in denial to accept that we are in an anthropogenic climate emergency while equally attending to people who are in denial that we have any chance of saving ourselves. To truly believe there is no hope is a sad situation for people whose environmental knowledge weighs heavily on their hearts. The gulf between the two groups of deniers could not be greater; it is a chasm that is difficult, if not virtually impossible, to cross. Thankfully, there are middle-roaders who are not in denial of the reality of a climate emergency and who are not in denial that the outlook for the human species is hopeless. These middle-road people believe life on Earth is worth fighting for, and together, they are doing their bit to bring about an education for change as best they can in areas within their influence. Hope is the energy that fires the middle-roaders to climate action both personally and collectively as a community of concerned citizens. They are the local environmental activists who, every Friday, stand on a street corner and accost the people passing by to join them in the dance for a safe environmental future.

Then there is the fact that climate change will continue to affect us all equally. This is not someone else's issue, like the people in undeveloped countries for instance. Most people feel incapable of taking unilateral action not because they don't want to but because they just don't know how to, and some don't want to be ridiculed as greenies or worse. In the case of the changing climate emergency, we are both bystanders and perpetrators of the crimes against creation. It is serious stuff to talk about – a very unpleasant and unnerving conversation, to say the least. I know this from first-hand experience of rejection for discussing climate change or environmental destruction.

There is a school of thought that argues that denial of anthropogenic climate change originates from a sense of despair regarding the seriousness of the consequences that will bring about an uninhabitable Earth and the possibility that it is too late to prevent it. It also comes from a sense of inadequacy to do anything about it if governments are not prepared to step up to the plate with real answers to alleviate their despair. One can be in empathy with them when there is no intention from some governments to change the status quo – that is, no intention of getting out of burning coal, gas, and oil in the next ten years. There is also the problem of how to change the consumerist culture that we are so immersed in. People who are in denial can express their confusion by denying that there is a problem, blaming people for promoting action, accusing people of stirring up trouble, outlawing people who are protesting for climate action, providing arguments that the science is not accurate, or criticising the science as fake news, amongst other responses. The plain fact of the matter is that environmental information comes in small bites of news on the mainstream media with little follow-up, so it can easily be dismissed. For example, during the Covid 19 dilemma the International Panel on Climate Change 2021 report was published but the media quickly reverted to their endless discussions on Covid and its attendant statistics. In less than a month, The IPCC Report has virtually gone off the air and under the radar of our collective consciousness. One climate scientist I listened to said it may well be the last report. Is that because what is in the report will be self-evident over the next six years?

As climate change intensifies, we can therefore anticipate that people will continue to collude in creating mechanisms of denial as a way of coping with reality. The tragedy is that when the worst of the changing climate comes to pass, we, as wise humans, may not be prepared to manage the fallout that will inevitably follow rising global emissions exacerbating global heating. The best we can do is be gentle with one another and respectful of one another's position. Climate scientists, along with well-informed environmental people, are already suffering from what they know and anticipate will happen to our societies. Their knowledge is already a heavy burden to them and one they reluctantly lay on us. No one ever wants to be the bearer of bad news; however, for better or worse, we are in this mortal dilemma today.

This is a huge challenge, and we can easily be daunted by it before starting out on our mission to turn global heating around, but when our mission is accomplished by stabilising the climate, we will look back and think our victory was inevitable because at the final clarion call, we responded with love for all life. Responding to the challenge of climate change is now a 'David and Goliath' situation, but we know who wins. So let us embrace one another for the journey, embrace our home planet, and embrace our biomes, ecosystems, and biodiversity, and together, we can help create a renewed wonderland and give our beautiful, fragile jewel hanging in space a good shine. The job may appear to be hugely complex, but the pathway is simple, so let's get on that path!

Tipping Points

Climate scientists are on point looking at possible tipping points. However, tipping points are elusive in that there are many unknowns as Mother Nature can always throw up something that has not been considered at this time. Tipping points may occur where the changing climate and environmental degradation could push any of Earth's ecological systems into abrupt change that could be irreversible. For example, grasslands under climate pressure can become deserts, or rainforests could become tipping points through drought, global warming, bushfires/wildfires, or deforestation. There are climate and ecosystem tipping points that can drastically influence how Earth functions as a land mass or the ocean; both are interlinked in their function.

Climate scientists point out that there are several tipping points that require our attention. There doesn't seem to be a consensus as to when these tipping points might turn into runaway climate disasters, but at least we have been alerted to the possibility that any one of the tipping points could happen with disastrous consequences. There are so many variables that could influence such a magnitude of change. This is where the precautionary principle can come into play – that is, better to know of the possibility than be taken by surprise. As with every ecosystem, a small change can have a huge effect (remember the flying foxes – just an increase of two degrees Celsius), and so it is important to understand the following tipping points that have been introduced earlier in this book:

1. Melting of permafrost on the tundra
2. Loss of ice sheets on West Antarctica
3. Loss of ice on Greenland in the Arctic
4. Amazon Forest deforestation (logging, drought, or dieback)
5. Boreal forest shift
6. Monsoon shifts
7. Coral reef die-off
8. Loss of biodiversity
9. Wildfires

A number of these tipping points were predicted a decade ago, and some are already active. They are visible and now able to be measured with a degree of accuracy because of advancements in technology, even to the extent of sonar and satellite imaging, which present even more reliability than the handheld camera, thermometer, or tape measure.

Words like 'trigger', 'unprecedented', 'mitigation', 'adaptation', 'positive feedback loops', 'crossing the threshold', 'unchartered waters', 'point of no return', 'runaway', 'incremental', 'cascade', 'guardrail', 'climate bomb', 'moving on to catastrophic', and 'apocalyptic' are all used to describe tipping points. The words alone are enough to cause a response from us forward-thinking people. When you look at the list of potential tipping points, it is not easy to dismiss them as some of them are already active and showing signs of breaking down or fragmenting at least. So what is known about the suggested tipping points?

(1) The tundra is melting, and permafrost is being exposed to more sunlight and is thawing at an unprecedented rate, releasing methane into the atmosphere. Methane is so much more potent than carbon dioxide as a greenhouse gas, hence a real threat to adding to the greenhouse effect.
(2) West Antarctica is losing glacial ice sheets. Ice sheets are glaciers that flow from land into the sea.
(3) Satellite imaging clearly shows that the ice sheets covering Greenland in the Arctic Circle are melting. Ice melt is pouring into the ocean; less ice means more dark water, and more dark water means more heating of the ocean and ice, which creates a vicious cycle.
(4) Again, the Amazon Rainforest is so large, the impacts can only be fully assessed by satellite images. Quite clearly, the Amazon is struggling with deforestation, drought, global warming, and wildfires.
(5) The boreal forest is experiencing climate change, especially global warming, that is hastening the thawing of permafrost, the early onset of logging, wildfires, and human interference by way of mining and hydroelectric

developments. The northern ecosystems of the world are heating faster than expected, and the most threatening issue is as permafrost thaws, excessive amounts of methane will be released into the atmosphere, which will exacerbate global warming. As coniferous trees are burnt by wildfires, deciduous trees are replacing them, which is a potent sign of the changing environmental landscape.

(6) Monsoons are driven by a thermal contrast between land and ocean and dependent on the availability of moisture in the atmosphere. The changing climate can affect both wind and rain. With ocean temperatures rising, the prediction is that the monsoons will result in heavier downpours and increase flooding, or a shift could occur in the monsoons, resulting in drought. Climate change induced by global warming could change the landfall of monsoons. At this stage, it is an unknown as to when there could be a tipping point that would alter monsoons, which are the source of life for millions of people and their crops.

(7) The tipping point for coral reefs has already been manifested in many corals around the world. The most important coral reef lies off the coast of Australia, known as the Great Barrier Reef. It is already showing real signs of stress, and it is well documented that most coral die-offs are the direct result of global warming. Coral die-offs mean that the dead corals are then exposed to invasive species, which continue to kill off the reefs, such as the crown-of-thorns starfish.

(8) I have added the loss of biodiversity because we have lost approximately half of all biodiversity and are continuing to lose biodiversity at an alarming rate because of global warming, human interference and exploitation. The human species needs everything in the biosphere to survive and flourish.

(9) I have also added bushfires/wildfires as an important consideration because 2020–2021 presented many

worldwide examples of the power of fire to decimate forests, landscapes, biodiversity, habitats, ecosystems, and homes. Many of the wildfires are caused by lightning strikes, which are a consequence of disturbances in the atmosphere because of global warming and the evaporation of water into the atmosphere. If we are to continue experiencing wildfires, mega fires, or giga fires in the near future, then the released carbon from the fires would exacerbate global heating, resulting in feedback loops or domino effects on ecosystems and biodiversity.

Since I am not a climate scientist, I do not want to expound too much on tipping points. It is, for me, a scary topic anyway. However, if climate scientists are still on edge about which tipping point will give in first, then we cannot know either. From my research, there are differing opinions. Some climate scientists have concluded that the melting Arctic is already active as a tipping point; others say it is too close to call. This is where my decades of following climate science kick in; I wonder, do they not want to say it is as bad as it is? For us, it is still important to register that it is we who have created this dilemma that is unnerving many climate scientists. Sadly, we are still perpetuating a lifestyle that may tip one or another of the biomes or major ecosystems into a state that is irreversible; the possibility is certainly well within our lifetime. This is an ongoing narrative at play and one that we can all observe from our lounge room couch as each news item brings scientific research into our homes and our minds. For an example of a newsflash, it has just rained in Greenland for the first time in recorded history. Did anyone hear that and wonder what it could mean for them? Another potent revelatory moment is when a news report is resplendent with photographic evidence of houses on the north-eastern coastline of Australia being battered with ocean waves, resulting in washing away the cliffs that the houses are sitting on. Ask yourself – is there any possibility that the houses can be saved? Can a deluge of rocks save the houses, or will the power of Mother Nature prevail? Just because it has gone off the news headlines doesn't mean the ocean has subsided and the crisis has been averted.

In conclusion, tipping points are defined as irreversible changes in the climate system and biosphere, and there are multiple reasons why these could happen, but increased global heating is in all the possible scenarios. Tipping points could become reality if the global temperature exceeds 1.8 degrees Celsius. We have known of the potential of catastrophic tipping points for twenty years, but insufficient action has been done globally to arrest the possibility of irreversible changes to the overall stability of the environment. What is confirmed about tipping points is that climate change and other anthropogenic activities are risking triggering changes, which could then trigger a cascading effect in the biosphere or atmosphere that could/will have huge consequences for humanity.

Societal Chaos

You need to close your eyes to read this brief explanation of potential societal chaos as a worst-case scenario. Climate science has been around at least since 1965. So for nearly sixty years, we have known that if we continue to pollute the atmosphere by releasing captured carbon back into the air, then it would be accumulative and impact on ecological and biological systems. We have been told that this would happen by climate scientists for decades, and the damning narrative has not changed; it has become more urgent each decade. It was 1987 when the first United Nations General Assembly agreed that climate change is real and that we are causing it; that was over thirty years ago, and what effective action has been taken to stem the tide of global warming? We think it is a 'pollution of the atmosphere' problem, but it is a political issue of inaction; for some reason, those in office have a psychological resistance to address it as urgent. Given the amount of time that has passed, we need to accept that a reformist approach is not going to work anytime soon. It must be a confrontational and revolutionary response. Hence, the Environmental Revolution is now well underway; all that is needed is you, your family, and your friends.

What then are the potential consequences of refusing, as a collective humanity, to act on climate science that clearly states that if we continue to elevate global warming, it would be our undoing? The consequences of our continued inaction are most alarming and deeply frightening, so harmful that it is difficult to commit to writing it on paper. No climate scientist wants to share this message, and neither do I, so it takes great courage to write about what our forecast is for the future. Our inaction may have huge ramifications, the possibility of societal collapse and subsequent chaos, and perhaps we are setting up the conditions for our own extinction as a species. This is a radical statement because even amongst many climate scientists who also hold that position, they are not prepared to come out so bluntly and say it, but it is there in the background of their thinking. It is the elephant in the room!

The plain fact is – as covered earlier in this book by examining biomes, ecosystems, and biodiversity – we could bring about our own demise because we are just big bugs on the Earth, members of the animal kingdom, and dependent on the health of the planet. When you study the current climate science as well as examine the mathematics, chemistry, physics, and photography of the changing climate, all of which state absolute facts, what is the alternative conclusion that you can come to? There are many scientists who are now pondering on what our future might look like if tipping points occur.

So what is our prognosis for the future of humanity? Some climate scientists are not afraid to speak out now as their special environmental issue is a 'tipping point' that is beginning to tip, like the melting Arctic. As time passes, the theory of possible societal collapse is gaining momentum and is being hotly debated by those concerned about escalating environmental degradation. It is quite clear that we need to imagine a new way of being in the world. We will need to engage in deep adaptation because our lifestyles in First World countries are unequivocally unsustainable. As a species, we are now in unchartered waters; we are devouring Earth's gifts and changing the chemistry of Earth at an unprecedented speed to the point where eventually, if we continue depleting the gifts of Earth, there will be nothing much left to support life and a very angry Mother Nature will be in control of the climate.

The thrust of an immediate need for intensive mitigation and profound adaptation to stem the tide of environmental collapse is powered by a timeframe – that is, less than ten years to adapt to a reformed way of living to stop global warming in its tracks and prevent runaway climate change. More potently, we are facing a climate-induced societal collapse. This will inevitably come about by the shortage of food because of global heating and most likely water as well. There is a consensus that food security is a big issue as already, there is a reduction in agriculture productivity in some countries, largely because of the changing climate. There are many stories of farmers around the world who are unable to plant because the seasons are changing and cattle producers who have had to sell off their livestock and sometimes 'put down' their herds due to unprecedented drought. There has already been a breakdown in the way we produce food; some have ended up in poverty and mental illness as farmers struggle with what they have never experienced on their farms over multiple generations of family ownership. While some farmers are accepting their fate and looking for alternative uses for their land, others are in denial and others are despairing that the good times are potentially gone for them.

The fact is that we cannot have continued growth forever; everything we think we own and control comes as a gift of Earth. The changing climate will disrupt every aspect of people's lives; there is no escape route, only mitigation, followed by adaptation. We can choose to continue as we are and end up with a wasteland over the next fifty years, or we can engage in a new pathway and, over the decades, restore Earth to a wonderland. We have all experienced the wonder, mystery, and magnificence of the natural world. We know how good it feels to be surrounded by nature. If this is what our hearts desire, then our heads must fall in line, and we must strive to create a new, better, loving paradigm in the future with consideration for the totality of Earth's community of creatures.

To engage in immediate mitigation towards deep adaptation, the big issues that will need to be addressed are food and finance as both will be unstable. Human society will have to adjust to food shortages as agricultural conditions are changing and meat production can no longer be sustained if we want to feed the entire population of people plus biodiversity, which we must protect. Financial institutions will struggle and possibly collapse; for example, insurance companies may not be able to continue to prop up insurance policies that are clearly hostage to the impacts of global warming. The question of house insurance has already arisen, with insurance policies being denied to people in potential bushfire or flooding areas. With the Arctic melting, sea levels are rising, which will cause flooding and will threaten land values, not to mention those who stand to lose their homes because of rising sea levels. There will be an increase in lawsuits and litigations; blame and fault will be in continual debate. The issue of millions of environmental refugees and mass migrations from low-lying cities and countries will have to be resolved. All of the above will lead to civil unrest and combative politics. Media will play a role one way or another – that is, a positive and helpful role or a negative role – and engage in the blame game, which has no place in any endeavour to bring about the required change in the way we live. The blame game has real potential for societal disruption and chaos. To anticipate societal collapse and potential chaos, it is vital to come up with a way forward for humanity.

To address the possibility of societal collapse, we need a new agenda. It requires serious soul searching to seek not only a revised respect for the natural world but also a very supportive respect for one another as we journey together.

- Continue to educate ourselves on the health of Earth
- Address our own environmental footprint
- Engage in our own ecological conversion
- Develop resilience to adjust to a new way of being in the world
- Think about what is most precious to us
- Be prepared to let go of some of the things that we love
- Be community centred
- Participate in community activities that will help restore local environs
- Allow Mother Nature to guide us in a renewed relationship with the natural world
- Immerse ourselves in nature so that we are empowered to defend nature
- Be prepared to engage in the great work of our time
- Be futuristic and imagine a world where we live in unison with nature
- Be people of hope and faith that we can rise above adversity

We will experience all kinds of emotions as we undergo the above. However, if we accept all of the above, then humanity can rise like a phoenix from the dust; we can be victorious. We can modify our consumerist lifestyles, change the way our economy operates, overcome our eco-anxiety, and then we can set a new course towards environmental sustainability for all. We can take solace that the best of what it is to be human will rise to the challenge. No one has the lead role in this endeavour; we are all novices because this is a road untraveled by humanity, and we will all be in unchartered waters with a dodgy compass. This is truly the great work of our time – that is, to restore Earth's ecosystems to their former health – and everyone can and must play their part if we are to overcome this extreme adversity.

This is not meant to be a gloom-and-doom narrative; we can rise to the challenge as evolutionary 'wise humans', and from potential death, we can have a rebirth without the dying. This is not meant to paralyse us into inaction; we need to go about our normal lives but with a new mindset – that is, to be respectful of the natural world that supports us and make good choices for the health of Earth. You can join those who believe in our future, that we can turn around the unravelling of our world and commit ourselves to envisioning a new and beautiful way of being earthlings.

For those of us who have struggled with knowing that we could end our evolutionary journey as we know it, we must gain the energy and emotional strength to sustain us always forward to action. We can develop our own sacred spirituality of Earth, our connectedness to the natural world, and accept graciously our potential destiny while finding space in our minds and hearts to allow the mystical to strengthen us. Modern technology has largely separated us from physical, emotional, and mental closeness to one another and the natural world. However, once we have the knowledge, it can be transformative in our lives and take us on a new journey to an even better and richer human existence.

The mantra for COVID-19 suppression and eradication is that 'we are all in this together' and that if we work together as a concerned and united community, we will come through it. This belief has worked before, and it could work again if more people wake up to what is ecologically important for humanity. It is not a time to feel sorry for ourselves; it is a time to be our most amazing wise selves, and experience has shown us that it can be done. Whenever and wherever there is a humanitarian crisis in the world, communities rally to give assistance to one another, even to strangers.

While the future does not look so bright, many millions of us are committed to environmental change, and we will continue to work towards that end. We welcome everyone to join the dance towards a renewed Earth and a new and wonderful way of being in the world. We have the intelligence, we have the vision, we have the skills, and we have the technology, but now we need the 'will' to step up to the plate and accept that we are in a bit of a mess, and we all have the 'power of one' to make a difference. There are multitudes of people who are exercising the 'power of one' to make a difference to the health of Earth, and we have a choice to tag along and assist wherever we can. Societal chaos need not be our human story. We can rethink 2050 to 2030 and envision a world completely embraced by nature, the biosphere that has allowed us to evolve as a species and thus restore ourselves as a wise species, full to the brim with ecological wisdom.

The message from all of the above is that we can have a future that can be blessed and radiant with life if we can just get the pathway straight. The pathway must be paved with reverence for the natural world, respect for all people no matter what their philosophical position, love for all creatures that have shared our evolutionary journey and ecosystems that support us, and learning to be resilient in head, heart, and hands in times when the going gets tough. We environmentally educated people are called to lead the way under a banner of love for all life.

CHAPTER NINE

Philosophy of Our Ecology

Today over 50 per cent of people live in cities, the human-built environment. Therefore, many people are separated from nature. This rapid change from being inextricably immersed in nature to being almost completely divorced from nature has impacted on the general well-being of people. It is not an issue that was immediately evident, but research has shown that the human species, for general health and a sense of well-being, needs to be in contact with the natural world, and it seems that people require more than just breathing air, looking at the sky, or experiencing rainfall. People experience a sense of peace and joy in their gardens, if they have one, and they enjoy going to the beach or holidaying in the country. Real estate is most expensive if it has a vista of the natural world. So why is that?

Everything Is Alive

During my school years, we were taught that there were some things that were alive and other things that had no life in them. This teaching has now been overturned by the scientific world because everything has life in it, and everything combines to produce suitable conditions for life on Earth. The whole Earth is a living organism – that is, Earth, as a planet, is alive. That means living things, organisms, interact with things previously thought of as non-living things, inorganic – that is, not alive – such as rocks, mountains, water, and air.

Organic and inorganic things work together in one complex system to provide conditions that enabled life on Earth as different from Mars. There are billions of planets in the universe, but so far, exploration of outer space has not revealed any other planet that has life on it. This, of course, makes life on Earth very, very special and needing thoughtful care. Life on Earth is a mystery, and scientists are continually exploring it to understand how Earth functions as a living organism. Every action has a reaction, so what we do, as one species, is important to the rest of the natural world. If we understand that Planet Earth is a living organism, then we know that anything living can die. Given the number of species extinctions because of human domination and climate change perpetuated by global warming of our making, Mother Earth has something to tell us, and we should be listening.

Ecopsychology

Ecopsychology is a comparatively new social science. It reflects on the fact that the natural world is not just the environment that we live in; it is also us, our whole human selves. Ecopsychology has a huge following now as psychologists make connections between planetary health and human mental health.

Eco-psychologists believe that it is psychologically and emotionally disturbing for humans to live disconnected from the natural world, so perhaps such a disconnect could contribute to a mental malaise or depression. People can suffer from depression, anxiety, or stress or display other symptoms that suggest all is not well with our society. Ecopsychology, therefore, is about healing the rift between individual people and their lack of connectedness to the natural world by engaging in therapeutic activities such as meditation and mindfulness, spending time in parks and gardens, or experiencing a nature immersion, such as 'going bush'. Bricks and mortar, concrete and asphalt, and closed-in apartments separate people from nature. Some people may go their whole lives and never touch dirt or plant anything. Ecopsychology is supported by research that reminds us of our ecological history – that is, we have within us a deep affinity with nature and a longing to be connected to nature. This phenomenon is called 'biophilia', a term used to describe our innate love for nature. There are times when we need to ditch technology to release our wild hearts and minds and be engaged with the natural world. Immersion in nature is good for physical, mental, and emotional health for a sense of well-being.

Biophilia, a term used by Erich Fromm in 1973 gives us an insight into the natural world and can bring us to a new place – that is, love for our home planet. To learn to love our home planet, it is important to understand our beginnings and evolution as a planetary people. As the environmental crisis continues and the natural world is more and more depleted, humans will suffer impoverishment spiritually, emotionally, and physically, and we will feel in our hearts that sense of loss even if we don't know where that feeling comes from or why we feel a malaise in our lives. We belong to the animal kingdom, and we are most at home in the natural world. For health and well-being, we need to reconnect with the natural world and plunge ourselves into the wonders of nature, which can be a great antidote to a depressed feeling in our lives or suffering from ecological anxiety.

With all the talk about global warming, climate change, and the loss of ecosystems and biodiversity, we can be forgiven for not thinking about how the environmental crisis dialogue affects our human spirit right down to our very souls. Have we lost our connectedness to the natural world so fully that we are, as a society, experiencing some form of insanity? In our daily routines, do we forget our intimate relationship with the natural world? Ecopsychology reminds us to open our minds, hearts, and souls to love our splendid, lavish, verdant, and very much alive planet. This is so important to our mental health and general well-being. Once we are awakened to this phenomenon of needing to be intimately connected to the natural world, we can begin to nourish our hearts and minds by immersing ourselves not only in the natural world but also in the story of our beginnings, the great cosmic story of us. Remember, everything on Earth is made of stardust; we are all made of stardust, really.

Cosmology

Cosmology is 'big picture' stuff for our minds to grapple with to understand our place in the cosmos, in the universe, and as earthlings. Cosmology is the study of the cosmos, that ever-expanding universe. As human consciousness evolved, people wondered about the visible created world and sought explanations for how it all came to be. Before the advent of modern cosmic science, early history abounded in myths and stories to explain the presence of the universe, Earth, and human existence. The big metaphysical questions challenged scientists to discover who we are as a species, where we came from, and what our purpose is. Scientists have looked beyond the stars with telescopes and satellites and into the microscopic world of atoms to discover answers.

It is difficult for us to conceptualise the great story of the universe. It is extremely challenging for us to understand the big bang theory and to know that we are part of everything that makes up the universe. As we came to know the 'great cosmic story', it changed everything humans believed for nearly two hundred thousand years. The biggest challenge of all is to see ourselves as interrelated and interconnected to everything we see. Everything is encompassed in this basic but momentous story of the birth of galaxies, stars, planets, and minerals. We are made from stardust, along with every creature that has shared in the evolution of life, just as we are carbon, air, water, and soil. What an awesome concept to get our minds around!

As cosmologists began to discover how our universe was created, new and amazing stories began to fill the void of the unknown about the existence of everything. It is important to note that we are the first people to know this amazing story, the cosmic story of creation, from a scientific perspective. This story is the greatest scientific story to be communicated to us and a story to rejoice in. It is a sublime narrative of an unfolding universe over fifteen billion years. It begins the advent of time and matter and us as members of the human species who are part of this awesome story.

The universe encompasses the entirety from what scientists call the big bang. I have always thought the 'big bang' was a strange title for such a momentous beginning of everything we know and experience. While science has unravelled much of the great story of the universe, one question remains a mystery – that is, what caused the big bang? However, what we do know, thanks to the scientific community, is that the cosmos is resplendent with millions of galaxies, billions of stars, and planets. The big bang that ignited the universe occurred about fifteen billion years ago. Over time, rocks, stardust, meteors, and debris from the big bang formed galaxies, planets, and stars. Planets came together after billions of years of bombardment of cosmic matter from space. During this time, our home planet formed slowly and violently as it was a burning ball of fire until it cooled around four billion years ago.

Cosmology then is a study of the mysteries of our beginnings. It is the greatest story ever told, for it invites us to see ourselves as part of this amazing story. It is the story of the universe, the story of our home planet, Earth, the story of the advent of *Homo sapiens*, and the story of us. Everything and everyone emanated from the big bang and is created from cosmic matter, made from the stuff of dead stars. Everything on Earth is interlinked and interwoven into the fabric of universal life. When we understand how the universe began and how awesome our home planet is among millions of other planets, surely, we can learn to love this beautiful, fragile jewel in the universe and, in loving her, take care of her.

Earth Ecology

Our home planet is the greatest natural wonder of creation because it has life on it. To love our home planet, we must first get to know her. So many of us just live day to day without ever stopping to consider the wonder of life that we enjoy.

Earth is a miracle. Life is a great mystery. So many elements had to come together – energy from the sun, water, and an atmosphere that was imbued with oxygen. All living, breathing creatures depend on oxygen in the air. Over time, soils formed from Earth's crust, and the decomposition of primitive life forms provided the building blocks for vegetation. All life is interdependent. Energy from the sun, water, air, and soil is part of who we are as Earthlings. We are Earth people, and our lives are totally entwined and coupled with the rest of the natural world. All life has a common ancestry, forming an extremely diverse community of beings in a symbiotic relationship, and we are integrally immersed in it.

Earth formed to its present expression about four and a half billion years ago. It took most of this time to generate conditions that enabled life to come into being. Everything evolved in perfect balance and in harmony. Many millions of life forms evolved and disappeared, such as the dinosaurs, which are iconic creatures, but there were many billions of other creatures that shared the journey of evolution before the human species entered the great story of life on Earth.

The Earth is alive, and its natural functioning generates ice ages, volcanoes, earthquakes, hot springs, tsunamis, wildfires, floods, and drought. Earth is a living organism. It has a life of its own, and it continues to express itself through its normal activities, which we sometimes call natural disasters. What is important in understanding Earth's ecology is that life is an astounding, marvellous, and extraordinary miracle, and we are part of this miraculous story. We just need to understand how Earth functions to preserve all the conditions that initiated life in the first instance. If we understand that life on Earth was a process, a slow process to establish ecosystems that now support an immense diversity of life, then perhaps we can reflect on the importance of keeping all ecology in perfect balance and harmony to allow life to continue and evolve over time. It took millions of years to make conditions on Earth perfect for human life to flourish, but we are now tipping the balance. The human species never thought that we could change the conditions for life on Earth, but we have, and it is measurable, so it is extremely relevant to grasp that our behaviour has consequences.

Human Ecology

Humans are comparatively new comers to the great narrative of life. Over two hundred thousand years ago, human life evolved and differentiated from that of our primate kin and gave birth to human ecology. *Eco* means 'home'; ecology is the study of 'home', meaning our home planet. The ecology of the human species evolved over many thousands of years, so it is important to bring to our minds that we have a common ancestry with all other creatures. We are interrelated with all of the natural world that share our evolutionary journey. If before, we thought we were a special, exclusive species, it is not so now; as scientists understand the DNA of humans and other animals, we know that we are more like other creatures than different from them. Humans, however, are very special and set apart from the rest of the natural world because we alone have the capacity to dream, to plan, to know our past, and to control our future. Because humans have been gifted with an amazing brain, we are the dominant species on Earth. Having said that, it is also important to understand and appreciate that humans are completely dependent on the rest of the natural world. Humans utterly depend on air and water and are nourished by food from plants and animals that live on plants.

Human ecology deeply connects us to the natural world by our evolution as a species. Today half the population of the world live in cities removed from the natural world, even though housing, infrastructure, and everything come initially from the other-than-human natural world. So while we are surrounded by natural elements, we, in general, do not reflect on this fact. However, there is a spiritual side to humans, an Earth spirituality that draws people to parks and gardens, holidaying in the country or by the sea. Humans want to travel and see the great natural wonders of the world, such as glaciers, coral reefs, deserts, rock formations, mountains, lakes, rivers, and especially forests. Therefore, some countries have set aside national parks as sacred, so there will always be places for people to experience the natural world in the wild. People also love their gardens, whether for aesthetic value, such as flowers, or to play in the dirt by growing their own vegetables.

Homo sapiens are interconnected to the natural world for food for the body as well as for spiritual nourishment. Humans are so much more than blood and bones; we have a spiritual side to us as part of who we are as a species. Human ecology is intimately linked to macro ecology, the totality of ecosystems. We may think we are separate from the natural world, but thinking that way does not change the reality of interrelatedness, interconnectedness, and interdependence. We know this to be true because we cannot live more than a few minutes without air, a few days without water, and not very long without food. We are totally dependent on the ecosystems that provide what we need to live. Once we are committed to this truth, we will be in the best mindset to protect and nurture all life as they are kin. When we really understand human ecology, we are in the position to view the world differently, to learn to love our awesome planet and be prepared to protect it and, if necessary, to defend the totality of its biosphere.

Ecological Self

We all have ecological selves. We are Earthlings. Our sense of place is our house, our street, our suburb, our city, our state, our country, and our planet. We are intimately entwined in the great story of the universe. We are deeply immersed in the amazing story of Planet Earth, our home planet. We are extraordinarily imbued in the evolution story of the human species. Each one of us has a past, a present, and a future story of connectedness to the natural world; consequently, characteristics of our bodies changed, such as height and skin colour, to what was best suited for survival. Different languages, stories, and cultures evolved in unison with relationships to the Earth and its biodiversity.

Our connection to the natural world begins in a single cell of the billions of cells in our bodies. Each cell requires air and water. The

Earth is two-thirds covered with water, but so are our bodies about two-thirds water. Our bodies provide habitats for millions of bacteria and other organisms both inside our bodies and on our skin. Our body is a mini ecosystem. We could not live without this awesome relationship. Every minute 'bacci' and 'orgi' do their jobs to keep us healthy; so it is that we are air, water, soil, and energy from the sun. Many humans today are lacking in vitamin D because as city dwellers, they are largely indoor people and lack exposure to the sun, the source of all energy for plant and animal life. We know in our hearts that awareness of our connectedness to the natural world is good for us. People spend enormous amounts of money to buy real estate that has a natural vista, such as overlooking the sea, beside a park, near a river or lake, or backing on to a national park. People take part in all kinds of sports that take them into natural terrain – for example, camping, hiking, rowing, skiing, hang-gliding, and fishing. When children are in parks, they often try to make a hut out of fallen branches and leaves. Children can let their wild side express itself when exposed to the natural world. We are very fortunate to have access to so many parks and gardens to spend some leisurely time with nature.

It is so important that we all realise that we have ecological selves. Some people know this and nourish their call to the wild by becoming part of nature when opportunity allows. Scientific research in ecological psychology has proved that humans are happiest when they are surrounded by nature. Also, people are healthiest in mind and body when immersed in nature. Just as we all have a naturalist intelligence that needs to be developed, so do we have an ecological self that responds to creation. What is mostly needed is that we recognise this call to nature and take time to ponder on the miracle and majesty of the natural world. We can all gaze at the clouds or the stars in the ever-expanding universe. We still have a remnant of the hunter and gatherer in us to call us to the wild and grow our eco-literacy. Every heart seeks to be imbued with the living spirit of Earth. Surely, we must still have enough hunter and gatherer spirit in ourselves to want to leave a footprint that does not impact on the health of the ecosystems that support our lives. This deep awareness of our innate connectedness to the natural world is termed 'deep ecology'. Deep ecology takes us to a new level of our ecological selves and environmental consciousness.

Deep Ecology

The concept of deep ecology is a radical way of thinking. There are many who believe that this is where we need to be to change our relationship from one of dominance over Earth and its biodiversity to one of equality, respect, and deep commitment to the restoration of all life. Why is this? It is because we have plundered the gifts of Earth for our own needs and wants over the last few decades to the point where our demands on the gifts of Earth are unsustainable. This is, of course, something that we have not reflected on, but we cannot go on using up the gifts of Earth as though there is no tomorrow in the belief that there will always be an abundance of life-supporting gifts. This is physically untenable for the Earth.

Deep ecology wants us to rethink about how we use the gifts of Earth, and thankfully, millions of people are doing just that. 'Reuse, refuse, recycle, reduce, and rethink' remind people to be considerate of how human activity impacts on the health of Earth. This is the Environmental Revolution. This is the Ecological Period. It is a time when we are really thinking about our relationship with our home planet and all the biodiversity that supports our life. Ecological science, concerned with facts and logic alone, cannot answer ethical questions about how we should live; for this, we need *ecological wisdom*, which is embedded in deep ecology. Deep ecology seeks to develop ecological wisdom by focusing on a deep experience of the natural world, serious questioning of how we are living in this world, and a profound commitment to live the ideology of deep ecology.

The ideology of deep ecology accepts the philosophy that

- all creatures have intrinsic value in themselves and must be free to flourish independent of the human species,
- all creatures contribute to the health and well-being of life on Earth,
- humans must only take from the natural world what is needed for their existence,
- the human population must be regulated to accommodate all other life,
- the plight of the non-human world is at risk because people are making disproportionate demands on them,
- the domination of the natural world by the human species must stop,
- the human species must examine its conscience and accept a standard of existence that reflects the security of the rest of creation, and
- we must downsize our footprint on the Earth so there is equality for all.

If humanity had embraced this ideology thirty years ago, we would not be trying to reconcile our actions that have brought about the depletion of over 50 per cent of the biosphere, and we would be well underway in living in a right relationship with the total Earth community of beings. The opposite to deep ecology is a dangerous ideology that must be addressed:

- All life only has value if it serves humanity.
- The human species is the most important, and all decisions regarding the environment must only consider the human species.

- Earth is abundant in its resources for the human species. Resources are there for the good of humanity.
- The human species can multiply without regard for other creatures' right to exist.
- All power goes to the human species to dominate the Earth and extract from the Earth whatever is needed for the human species.
- We must increase the quality of life for the human species.
- Governments have absolute control of the resources on land, in the ocean, and under the ground.
- No challenge is acceptable to the use and abuse of the resources of Earth.

The deep ecology philosophy was available over three decades ago, but to a large extent, we did not heed the call to a deep ecology mentality paradigm. Today we are confronted with all of the above under the banner of 'business as usual', which we are, in general, addicted to. 'Progress at any cost' is the common mantra, and we are all guilty of engaging in this philosophy. Can it be that we are brainwashed into thinking the gifts of Earth are infinite, endless in their gift giving, or are we just plain ignorant, and refuse to accept that there will be a day of reckoning with the other-than-human natural world?

Deep ecology calls us to reflect on our personal and communal relationships with ecosystems and biodiversity. We can examine ourselves to understand the depth of our alienation from the natural world. It is this alienation and sense of separation from the natural world that has empowered our dominance over the other-than-human natural world. Deep ecology challenges us to think of our evolutionary history and the bonds we have to the rest of creation. We must arrive at the understanding that when we defend a rainforest, we are defending not only the rainforest but also our own future. At times, we need to think of our kinship with creation and take on the persona of a river, a bird, or a glacier and tell their story. If the non-human world was able to speak, what would it be saying to us from the abundance of their wisdom? Unless we really understand our true relationship with the natural world, we will struggle to have empathy with creation and feel nothing – no remorse, no sense of loss – as a forest or coral reef is destroyed or when we hear that the last white rhinoceros has died. There is only one female white rhinoceros left, so this tragedy will happen in our lifetime.

Deep ecology requires a monumental paradigm shift in understanding our relationship with the natural world. Deep ecology embraces the natural world as kin, as family, as a community of beings that share the gifts of Earth in common. Remember, we belong to the animal kingdom; we are just big bugs in the ecosphere. We must be about integral ecology, an ecology that places us firmly within the community of life of all beings. We are integrally involved in every action and response in the natural world; we are nature. Once we understand our personal connection to the natural world, we can then extend our thinking to the multiplicity and complexity of biomes, ecosystems, and biodiversity that are us; we are interlinked with all creation as there is no separateness in the natural world.

CHAPTER TEN

Assessing Change in Our Relationship with Nature

Over the last two hundred thousand years since humans evolved, there have been monumental changes in the way we have lived on Earth. Our relationship with the natural world has changed dramatically over the last fifty years. Each one of us is leaving an ecological footprint on the Earth which we are only just beginning to assess. The most significant footprint is the amount of greenhouse gases we have pumped into the atmosphere, affecting the climate and the biosphere. The fallout of our behaviour is evidenced in environmental degradation; however, we are in a new era, the Ecological Period, of awakening to the needs of the Earth and becoming more environmentally conscious, and so there are thoughtful people who are giving a very strong message that we must engage in environmental conservation as we cannot continue on the pathway of using up the resources/gifts of Earth as we have in the past. Quite clearly, physics, chemistry, and mathematics demonstrate that our current lifestyles are unsustainable.

Ecological Footprint

For those of us who live in developed countries, our ecological footprint is a clodhopper. We all have ecological footprints that are measured by how much demand we each make on Earth's gifts and its ecosystems. Some people in poor countries or undeveloped countries have very light ecological footprints because they do not have equal access to 'resources' (gifts of Earth) that people in First World countries enjoy. On the other hand, people in developed countries, such as Australia, have very large ecological footprints as Australia is a very large country with few people compared to other countries. If we compare our lifestyle with that of much poorer people, we really do enjoy the lion's share, and our footprint is a size six Earth's clodhopper. Six Earths would be required to sustain us if everyone in the world lived with access to as many gifts of Earth as me.

Housing is a good example of an ecological footprint. Many people around the world live in very small houses, some with just one room with no electricity, no tap water, and no appliances, refrigerators, or heating and cooling. In Australia, we generally live in houses with many rooms and shower daily, with a limitless supply of electricity, an endless supply of food, all manner of appliances and gadgets, a room even for our cars, and our very own movie theatre or gym. The ecological footprint of the 'one-room people' is very light, and they live well within Earth's capacity to support them. The average Australian lives what is commonly called an unsustainable lifestyle as we are eating into Earth's capital at a greater rate than ecosystems can regenerate. Governments talk about our lifestyles as sustainable if we recycle, reduce, and reuse, but the bigger question is 'What is unsustainable?' Quite clearly, mathematics would pronounce our lifestyles as unsustainable because we need and use an excessive amount of Earth's capital daily which we in general, do not reflect on.

So how is our ecological footprint measured? Those responsible for working this out say that it is our impact on the natural environment – how much farmland is required to produce our food, how much forest we use to build our homes and other structures, how many fish and how much meat we eat, how much fruit and vegetables we consume, how much water we each need to drink, clean, and grow our food, how much energy we need to run our homes and cars, and how much waste and pollution we each produce. Everything we have and use comes from the other-than-human natural world. Some resources are infinite (that is, renewable), but other resources are finite (that is, non-renewable), and so our environmentally unsustainable footprint is a very important consideration for all of us.

Ecosystems are extremely busy recycling these days, and we do not think about what they do and how they are coping with the ever-increasing demands on them for their services. For example, humans cannot do what the forests do in cleaning and recycling the air we need to breathe, but there is only so much carbon that trees can sequestrate before they have had enough. The CO_2 that trees cannot cope with is released back into the atmosphere.

Environmental education is very much about understanding the need for environmental sustainability. There are thousands of environmental groups around the world responding to the need to rethink our lifestyles and what we can do to make our ecological footprint smaller and therefore less challenging for nature to cope with us. We can still enjoy the wonderful life we are accustomed to, but we must be a whole lot smarter about the way we live. People today are much more environmentally conscious and are happily adapting their lives to a more Earth-conscious way of living. They are making good choices such as less waste, plastic-free living, less pollution, thoughtful shopping, buying locally produced produce, installing solar panels, choosing public transport, shorter showers, being electricity energy conscious, gardening, picking up after their pets, and defending the rights of nature. If we are serious about ecological conversion and addressing our 'unsustainable' lifestyles, then we can examine our ecological footprint and take the steps needed to ensure environmental sustainability for the total Earth community.

There is a Chinese proverb that says, 'A journey of a thousand miles begins with the first step, and a massive waterfall begins with the first drop of water.' Each one of us can be a step or a drop towards putting an end to our individual unsustainable lifestyle. Our lifestyles are well within our control, and we can audit our own clodhoppers on the Earth and reduce our personal size so that it fits well within ecosystems' ability to service us. Many people I know are doing this, and they feel much better; they have a sense of mental and physical well-being because they have chosen to divest themselves of the clutter of stuff to walk more lightly on the Earth.

Environmental Sustainability

Every society, institution, or business is talking about environmental sustainability, but what does that mean? Environmental sustainability has been the 'catch word' or 'war cry' for environmental action for decades. The problem as I see it is that 'sustainability' has been about sustaining ecosystems for the health of the human species; it is preoccupied with human posterity. Environmental sustainability that is worth thinking about has to do with the viability of all of the biosphere. Education on environmental sustainability is now a necessity for life. It must be worldwide and lifelong if we are to continue to enjoy a tomorrow. It challenges all of us to get involved with self-education so that we can be environmentally judicious, to read the signs of the times with a commitment to understanding how

Earth functions and what can be done to keep our home planet healthy for today and the future generations of all creatures.

The goal of sustainable development is to develop programs and products with the cost to Earth as well as the cost to future generations in mind. So to get a good grip on what environmental sustainability is all about, the first question to ask is 'What kind of lifestyle is *unsustainable*?' This is not easy to do. None of us feel happy to forfeit some, if not much, of the 'good life' that we enjoy, but it must be done. Doing the right thing for the health of Earth does not necessarily mean life will be less blessed. There is no dichotomy between the health of Earth and human health. Both are intertwined and interlinked, so we must get smart about the choices we make as individuals and as a society; the big question is who will lead us as earthlings into the Ecological Period?

This movement requires a psychological shift. The human consumerist philosophy is that 'more is better', but a quick study of the health of ecosystems demonstrates that this mode of thinking cannot be sustained. Environmental sustainability is all about 'enough for everyone forever', and so we need to educate ourselves on what it is that we really need and what each ecosystem contributes to a healthy planet if we are going to live sustainably; there is no dichotomy between humans and the rest of the natural world. We are called to an ecological conversion so that we are armed with knowledge, wisdom, a vision, and a mission to protect life-giving and life-sustaining ecosystems. It is said that necessity is the mother of invention, and we can be quietly confident that we *Homo sapiens*, 'wise humans', will innovate, invent, and create with the speed of light as the needs arise.

A good example is that electric cars/vehicles powered by the sun are on the horizon for everyone. When I recall that my parents, after my birth, brought me home from hospital in a horse and cart, their means of transport, and think about my means of transport today, I can easily envision a new way of commuting that has the health of Earth in mind – no problem. Every day our minds are blown away with human inventions in so many different fields of endeavour. As for problem solving, we excel at that too. You only need to listen to ordinary, everyday farmers and entrepreneurs who have suffered huge financial losses through bushfires, droughts, floods, and trade disasters to see what amazing and inspirational attitudes and skills people have brought to bear not only to survive but also to flourish and make good out of extreme disappointments.

Environmental sustainability – that is, the ability to sustain the health of the environment – is in our hands. The 'progress at any cost' mentality must be seriously addressed and action taken now to make the changes necessary to sustain all life in perpetuity, and that requires some serious soul searching and decision making. What are we prepared to give up in our lives to ensure environmental sustainability for all planetary life? Are we prepared to vote out of office our favourite political party because they want to spend billions of dollars supporting fossil fuels when they could spend billions of dollars securing our ecological future and jobs by supporting the research, growth, and development of alternative ways of doing everything? That is a hard call to make, but it is one that we must consider if we are going to speak for a healthy home planet, and we have no time to lose. We must bite the bullet, as they say, and vote out any government that is not prepared to take climate science seriously or vote in any government that is prepared to listen to climate scientists to get the job done.

Necessity is the mother of invention, and those who understand the crucial need to exit fossil fuels are pulling out all the stops to design sources of renewable energy for all mechanisation, including air flight. The epic around-the-world air flight of the

Solar Impulse was akin to the Wright Brothers' first flight by a mechanised engine powered by fossil fuels. Currently, aeroplane travel produces 3 per cent of air pollutants, so aeroplane flights are not sustainable in the long term. That is not a happy thought as we love our air travel! As we were born into a coal, oil, and gas era, there is still a strong resistance to making the switch, even though it is the only safe pathway for the future. Fossil fuels will inevitably go the way of so many other good ideas, but will it be too late? That is the big question. Climate scientists, glaciers, oceans, insects, and birds would say 'yes', it may well be too late.

Environmental degradation is part of the Earth and human history now. No one person is to blame; the plain fact of the matter is that life was and is amazing, but sadly, we do not realise that a 'good life' for the human species comes at a monumental cost to the health of the rest of the biosphere. We have effectively walked roughshod over the Earth's gifts, and every ecosystem has been degraded by human activity one way or another. We thought we could use and abuse the resources of Earth at will without ever contemplating the long-term consequences of our behaviour. Can you think of any ecosystem in close proximity to you that is still in its pristine state? It would be awesome, in the spirit of globalisation, if those of us who depend on the services of rainforests could do some sort of pecuniary deal so that the rainforests remain intact. Some environmental organisations are endeavouring to buy forests, especially forests that are home to endangered species. This is an awesome way to preserve forestry biodiversity for future generations.

Environmental consciousness is a good news story; in response to the environmental crisis emergency, there is a revolutionary rise in environmental consciousness. All around the world, people are becoming environmentally conscious. Media education through news reports about something unusual in the natural world and ecological documentaries are generating environmental consciousness for those who are willing to learn. With access to so much real-time data, it is possible to make connections as each piece of environmental information flows our way. For example, on the news cycle, we have been alerted to the plight of the Australian platypus, an iconic and extremely fascinating creature. Scientific studies on the illusive platypus inform us that platypuses are now considered to be a threatened species. The platypuses' survival status is due to the loss of habitats, polluted streams, water diversion, droughts, and global warming, so their watery habitats are at risk of drying up. The clearing of land, bushfires, and invasive species such as wild dogs and cats are also impacting on their survival. City slickers would have no idea about the issues platypuses have to deal with to survive as they are far removed from the built environment. If the creeks dry up through global warming, then there is no salvation for this iconic creature. Out of sight, out of mind, do we care if this really odd but amazing little creature goes missing from the natural world? The tragedy is that we most likely will not know until years after they have ended their evolutionary journey with us.

People have been encouraged to begin to develop their environmental consciousness with environmental awareness – that is, reduce, reuse, and recycle everything. Recently, another R was added – that is, *rethink*. This is valuable progress because thinking is what humans do best, and to think about our consumerist habits is a very good start for environmental sustainability. This level of environmental consciousness is usually personal and requires an individual response. It asks us to think about our own contribution to the environmental crisis and what we can do as individuals to conserve and preserve the gifts of Earth.

The next step towards environmental consciousness is to look beyond oneself and out into the wider world community. People are thankfully beginning to tie environmental information together to make sense of all the talk about the environmental crisis. Education on environmental consciousness is a powerful motivator for rethinking how to sustain all of nature. Government environmental sustainability policies are designed not only in response to scientific evidence but also from the popular wishes of the people. People power is very effective in defending wildlife and the natural world.

Earth literacy informs environmental consciousness so that we can move forward to sustainability. It enlightens us so that we can read Earth's vital signs and raise environmental consciousness, which inspires and generates environmental action. Environmental consciousness moves beyond the self into a new paradigm where we accept that we are just 'big bugs', members of the animal kingdom on the Earth and in the same environmental situation and predicament as all animate life. A collective love for our home planet will save the day. Change is in the air, and we can all be part of that change if we choose to accept our responsibility to address environmental issues around us in any which way we can, even if it is just a tweet.

Precautionary Principle

What is the precautionary principle? A basic understanding is that it is better to anticipate a worst-case scenario than to pretend something is not going to happen and to act accordingly. For example, it is better to anticipate the possibility of a child drowning in a home pool, so the precautionary principle would say, 'Fence in a home swimming pool.'

The precautionary principle is an extremely relevant principle for today as it was when it was first conceived as a safe way forward when considering 'business as usual'. As a result of this principle, environmental assessments and protections have been put in place, and so major developments are, in general, required to submit environmental impact assessments. Such assessments are vital to protect the non-human natural world.

The precautionary principle is especially applicable when an identified threat of serious or irreversible damage to the environment is identified. In such cases, the lack of a full scientific report on a given environmental issue should not be allowed to influence a decision on a project that could be potentially or most probably in conflict with the health of an ecosystem or its biodiversity. Conflicts of interest are real, and often the stakes are very high – 'progress at any cost' versus 'environmental catastrophe'. There are some environmental projects that fly through to the keeper because of loopholes in what is required for a particular development to take place. However, the precautionary principle is, in general, in play and helps decision makers to adopt measures, dependent on the best science, that are in the interest of the health of Earth and therefore its people and biodiversity.

Environmental scientists and passionate environmentalists may not always get the precautionary principle right as sometimes there are strong reasons to do something for the environment and equally strong reasons that something could be done that may not seriously impact on the environment. An example of this dilemma is bushfires. Traditionally, firefighters have had controlled burns to eliminate undergrowth considered to be a fire risk based on scientific forestry best practice. Some environmentalists oppose controlled burning as it is destructive of undergrowth and a threat to biodiversity that live in it. However, in the heat of the moment, when bushfires are fuelled by undergrowth and energised because of sustained drought, the intensity of the bushfires kills everything in its path. A vital discussion around bushfires has to focus on best practice when it comes to preventing or managing bushfires that are so destructive to the human and non-human world. In some situations, there are no winners when it comes to the execution of the precautionary principle as everyone and every creature suffers a loss when bushfires ravage forests. However, the precautionary principle provides an important process for decision making when it comes to all projects that have the potential to impact negatively on the natural world. Long may the precautionary principle reign, and may we be its astute guardians!

CHAPTER ELEVEN

Issues Worth a Good Look

Cradle to Cradle

It is inevitable that if you have got this far in reading this book, you will be wondering what can be done as a way forward. The cradle-to-cradle philosophy developed by Professor Michael Braungart and William McDonough in 2001 which focuses on recycling, opened a floodgate of innovative ideas as to how we use the gifts of Earth and retain the lifestyle to which we are accustomed while supporting progress as usual. This model offers a pathway forward, and it has been introduced in several countries as it is a wise decision-making process now and in the long term.

The cradle-to-cradle philosophy is in response to recognising that we have a terminal ecological footprint. It challenges a traditional cradle-to-grave philosophy and practice that has flourished in a consumerist, disposable, and materialistic society. Landfills are required to receive huge amounts of what is classified as waste. The cradle-to-cradle philosophy is a no-waste system. This philosophy recognises that Earth's gifts are finite in general and therefore must be recycled as many times as possible. It also challenges designers, fabricators, and producers of all kinds to create stuff that can be remade or recycled in some way.

'Cradle to cradle' seeks to address environmental sustainability in a very real way because it first recognises what is unsustainable and then proceeds to a 'think tank' response where new ways of working with nature become the topic of the day. The question is how do we reduce our footprint on the Earth and still have an amazing lifestyle?' 'Cradle to cradle' requires thinkers and inventors to rethink and reframe sustainable uses of Earth's gifts. Ecologists believe that we must study the natural world and relearn ways of recycling in conjunction with nature as no other creature requires a landfill.

If our ecological footprint must be curbed, then the cradle-to-cradle philosophy argues that it can be done in such a way that is effective in bringing about a richer, healthier life for all and an equally prosperous country. The implementation of the cradle-to-cradle philosophy will inevitably be introduced as the world runs out of resources. The best time to initiate a new way of being in the world is while there is still time to make a full but painless transition to a cradle-to-cradle system.

What are some of the ideas to be considered when implementing a cradle-to-cradle paradigm shift?

- Be conscious of the finiteness of many resources/gifts.
- Develop a respect for the role that ecosystems and biodiversity have in maintaining the health of Earth.
- Only take from the gifts of Earth what is necessary.
- Learn from the natural world how to recycle everything.
- Support the restoration of nature wherever it is possible.
- Look at what is unsustainable in our lifestyle and make changes wherever it is possible.
- Only produce goods that can be returned to the Earth, a no waste paradigm.
- Stop practices that have detrimental consequences to the health of soil, water, and the atmosphere.

- Reassess 'business as usual' practices to ascertain whether they can be more accommodating to the health of Earth.
- Focus on the conservation rather than domination of Earth's resources/gifts.

In summary, the cradle-to-cradle philosophy involves a technological and biological approach to completely do away with landfills; that is the challenge, and when we do not need landfills, we will know we have accomplished the cradle-to-cradle way of being in the right relationship with the natural world. As the population of the Earth increases, so will the demand on natural resources, and most of the resources of Earth are finite. We have to begin a circular economy where everything has added value by recycling – that is, everything made must be able to be remade into something else. The 'take, create, dispose, or destroy' system is constantly eating into finite resources, which is unsustainable. Rethinking, redesigning, and recreating everything must become a new paradigm for a cyclical model economy, or resource scarcity will become the new norm, which will have huge consequences for society.

The cradle-to-cradle approach must replace the current cradle-to-grave economy, or over time, we will have few resources/gifts from the Earth left. This is not a matter of faith. It is physics and mathematics, plain and simple, as Planet Earth is a closed system; there are no more resources after we use what we currently have for seven billion people plus all biodiversity that also depend on the gifts of Earth.

Carbon Sinks

I have referred to carbon sinks in relation to ecosystems, but we need to get our heads around the role of carbon sinks. Carbon sinks are nature's way of processing pollutants in the atmosphere.

Carbon dioxide is an important gas in the atmosphere. It currently makes up .04 per cent of the atmosphere, but today it is a hot topic because the amount of CO_2 in the air is rising; 70 per cent of global warming is attributed to a rise in carbon dioxide in the atmosphere, and it is the human species that is generating it. So how does the atmosphere cope with such a huge change in the amount of CO_2 belching into the atmosphere from our activities? Well, there are carbon sinks.

Carbon sinks are a natural phenomenon. A carbon sink is anything in nature that absorbs more carbon from the atmosphere than it releases into the atmosphere. The process through which carbon dioxide is removed from the atmosphere is called *carbon capture* or *sequestration*. Carbon is stored in forests of trees, plants, soil, the ocean, and the atmosphere and is continually moving through different ecosystems. Fossil fuels stored under the Earth are captured carbon sinks until they are released and burnt back into the atmosphere. Burning fossil fuels has caused what is called a *greenhouse effect* because more carbon is released than natural carbon sinks can process. Mother Nature also exudes carbon and very toxic pollutants into the atmosphere when she is about her volcano business. We are more influential than modern-day volcanic action when it comes to producing carbon dioxide from our fossil-fuelled lives. The result is that carbon held in the atmosphere is now something like 30 per cent higher since the Industrial Revolution.

So what is happening with nature's carbon sinks? An analogy would be that if you drink a glass of wine, that's fine, but if you scull a litre bottle of wine, you might not feel so good. So it is with carbon sinks. They have been doing their job, processing and cycling carbon around ecosystems, for eons. Now carbon sinks are being required to do this at a much greater rate; carbon sinks have to absorb a whole lot more than they can process and release it back into the atmosphere, which is more than what evolution had determined as their job. In the case of the ocean, this is now a real crisis because oceans are absorbing far more pollutants from the atmosphere and therefore becoming more acidic, which, in turn, has a domino effect on creatures that call the ocean home.

Just to recap on carbon sinks, the ocean is the largest carbon sink, absorbing a quarter of fossil fuel emissions. The ocean is suffering from acidification resulting from toxic acid rain, and so the chemistry of the ocean is changing. This is why scientists are always testing ocean water to measure its toxicity. There are other ecosystems that capture carbon dioxide, such as soil and all kinds of vegetation. Soils capture and hold a large amount of carbon dioxide, as does vegetation. Such biomes as the taiga and the tundra hold enormous amounts of carbon dioxide and methane, a very dangerous, potent pollutant, especially held in the peat bogs. Forests are also carbon dioxide sinks and store an incredible amount of carbon. This explains why bushfires are so bad. Not only do they destroy the trees, vegetation, and habitats of millions of creatures, but also, as they burn, they release their captured carbon back into the atmosphere. Again, you can see that nature plays a vital role in keeping Earth healthy.

In 2018, climate scientists (98 per cent of them) determined that we had ten years to turn global emissions of carbon dioxide and other greenhouse gases around; that leaves us with less than eight years, so talk of 2050 zero emissions may well miss the clarion call of Mother Nature. We just have to hope that the climate scientists have got it wrong, and we still have until 2050, but you can do your own research on that data. However, there are already very clear signs that climate scientists may well be correct. Just this week alone, wildfires are raging in Greece and Canada forests, so the burning will add to the carbon dioxide in the air.

On a more hopeful note, most governments around the world are committed to greenhouse emission reductions but with questionable success as emissions continue to rise at an alarming speed. There are carbon taxes, carbon emission trading schemes, and carbon offsets and incentives to advance alternative sources of energy. Some countries are really committed to cutting greenhouse emissions with considerable determination, which is a sign of hope for the rest of us who are waiting in the wings for our governments to switch to a pathway that embraces renewable energy to take the pressure off natural carbon sinks. Natural carbon sinks do an awesome job, but they are now stressed to the max, so it is up to us to stop greenhouse emissions to give them some relief.

Advent of Plastic

Plastic is another huge environmental emergency that has emerged as a 'must fix and fast' issue. You may think plastics have been around forever, but no; plastic, as in everyday use, is a comparatively new invention. From no plastic for me as a kid, I am now living in a plastic world, and that cannot be good for the health and well-being of people and all creatures. Plastics are a by-product of oil, a fossil fuel rich in carbon. Plastic items are found everywhere, and most household items, toys, and school and office items are made from plastic or have large plastic components. Plastic is extremely versatile, from very thin cling wrap to thick planks for building constructions, because it can be moulded into anything, from toothpicks to buildings.

So what is the major problem with plastic? Plastic does not break down easily. Everything on Earth is connected, and one creature's waste is another creature's delight, but nothing seems to want to recycle plastic in the natural world, and it does not decay like other materials. Plastic just breaks down into the tiniest of pieces, microplastics that will remain in ecosystems possibly for hundreds of years. It is not known how long it takes for plastic to be consumed or absorbed and returned to nature as it is not biodegradable. We bought into the myth that plastics are the saviour of other materials that come from the gifts of Earth, but the result is devastating to the health of planetary life.

Plastic has made an incredible impact on everyday life, and much of it has added great convenience to our lives, but it has come at a huge cost to the health of ecosystems. Landfills are the recipients of mountains of plastic. In some countries, piles of plastic products such as computers are burnt, returning the captured carbon to the atmosphere, adding to air pollution. Worse than that, people have been extremely careless and thoughtless in the disposal of plastic products as over the years, tonnes and tonnes of plastic items have been disposed of by being tipped into the sea. Thousands of ships have dumped their plastic rubbish overboard, while some countries have filled up barges with plastic rubbish and sent it out to be thrown overboard into the sea – that 'out of sight, out of mind' philosophy. Marine scientists who have explored the ocean floor say there is not a square metre that is not contaminated by human debris, namely plastic.

Rivers, streams, and storm water pipes have all carried plastics into the ocean, especially when there are storms and flooding. People have left their drink bottles and wrappings on the beaches around the world to be washed out to sea on the next tide. It is well documented by scientists that some of the plastic is breaking down into tiny pieces and floating in the ocean; it is called plastic soup. It is impossible to sieve the ocean and clean it of plastic debris. This is a human-induced disaster for the wildlife of the ocean and sea birds that live off fish. No creature can digest plastic. If it does not pass through the digestive system, it remains in the guts of creatures that will eventually die from starvation because their stomachs are lined with plastic bags and other plastic objects they think are food. How many dead birds are found and in their decayed bodies are a myriad of plastic bits and bobs, even cigarette lighters. Plastic in the ocean provides a chain of disaster as each creature eats another in the food chain, and humans are often at the top of the food chain. Scientists have even found microplastics in plankton, the bottom of the food chain. Plastic was an amazing invention and provided an incredible amount of exciting stuff for us, but it has created huge environmental problems for nature. On land, plastics have traditionally been buried and therefore out of sight. This is not so for the wildlife of the sea. It is very much in sight and dangerous.

So what is the problem with plastic? Plastic is a by-product of oil, so it has been a windfall for oil magnates, and we have been seduced by the convenience of plastic. Supermarkets are loaded with plastic wrappings; some items can have as many as three layers of plastic coverings. Plastic bottles and straws are a particularly bad use of plastic and totally unnecessary in countries that have perfectly good water freely flowing from their own taps. People are drawn into the convenience of a plastic bottle that will outlive them after it is used if it is not recycled.

In plastic consumerist countries, people in general are not aware of how much plastic is washed up on the shore because people are paid to keep beaches clean. I didn't realise this until I spent a couple of hours on St Kilda Beach, waiting with my son for him to board the Spirit of Tasmania. The plastic in the waves was horrendous, and if it wasn't for COVID-19 kicking in, I would have gone back there out of a bad conscience and clean up what I could. The beach itself had been cleaned; it looked amazing, as it always does, but below the surface is the plastic of death for many creatures of the sea.

My generation of motherhood used cloth nappies; however, modern mothers are seduced by the ease, convenience, and softness of disposable nappies. Disposable nappies have plastic components, and a baby will wear at least six nappies a day for over two years, which equates to more than 4,380 nappies, which find their way to landfills, and realistically, the plastic nappy will outlive the child that wore it. Multiply 4,380 times by the births in any country, and you get mountains of plastic nappies. To add insult to injury for the Earth, sometimes they are burnt in the landfill, sending more pollution into the atmosphere.

There is not enough space in this book to expound on the health issues humans experience from consuming plastic. We are beguiled by the convenience of plastic, to the detriment of our health. It is only recently that plastic in the food chain has been identified as impacting on the health of the human species. However, there is hope, true to form, in Australia; after some countries refused to take our plastic waste, entrepreneurial people are addressing the issue of plastic waste and pollution in many innovative ways. There is no limit to our genius in recovering plastic waste, even on farms that use heaps of plastic sacks and plastic bottles. Necessity is the mother of invention, and the necessity to deal with our own plastic waste has recycled plastic, created jobs, cleaned up the environment, and produced reusable plastic products. This is a 'do good, feel good' action, and everyone is a beneficiary. This is the cradle-to-cradle philosophy in practice, and given the green light of approval, this can be replicated with all waste plastic products over and over. You might like to google 'Boyan Slat: The Ocean Cleanup' to see what the 'power of one' can do to make a difference. He and his team are working towards cleaning up the ocean, no less. He is already sieving plastic out of rivers at their mouths before it flows to the sea. His work turns a 'bad news' story of rivers delivering plastic to the sea into a 'good news' story. Equally, it is so encouraging that towns and cities are stopping the use of single-use plastic bags. Fifty years ago, we lived without the convenience of plastic, so we can do it again. It is up to us to put pressure on our supermarkets to delete plastic packaging from their products as many have now done away with single use plastic shopping bags and life has gone on, people have adjusted to bringing their own bags. With thought and determination we can make the switch for the sake of a healthier biosystem.

Population Explosion

There is a powerful argument that the exponential growth in the number of human species is to blame for the environmental crisis. Contrary to some opinions, the exponential increase in population is not to blame for the burden on Earth's gifts. At the end of World War II, the population was around three billion people. Today it is more than seven billion people with a statistical projection that it will increase to nine billion before levelling out. This would not be a problem as Earth is generous in her bounty, but it comes with a big 'but' – that is, we constantly hear of the amount of waste. Waste is the real problem; our consumerism philosophy is at fault as we are serial wasters with little appetite for securing an equality of lifestyles for all people on the planet.

There is thankfully some respite on the horizon as many people in developed countries are choosing to have only one or two children, which means they will not be replacing themselves in the overall count because of illness or accidents. In Australia, the fertility rate is about 1.66 births per woman. This is a downward trend and possibly by design as couples decide to have less children; however, there is inconclusive but current research on the potential causes of infertility in the human species because of lifestyle and chemical pollutants in ecosystems we access in our lives. There is research done that argues infertility in couples is increasing, and there must be an explanation as to why that is so.

Infertility is on the rise in the other-than-human world – for examples, animals, fish, and reptiles, such as turtles – and we are left to wonder why. Live births in turtles and crocodiles are currently being researched, and early assessments suggest that because of global warming, there are considerable changes in the ratio between males and females being hatched. Early research findings are that females are hatching in far greater numbers than males. This could have implications for the human species as infertility is on the rise in men. Infertility in the human/animal kingdom may also be impacted by global warming, the changing climate, and, more specifically, pollution via chemicals in the atmosphere, water, soil, and the food we eat. Chemicals designed and used to destroy other-than-human biodiversity may eventually impact on our ability to reproduce if it is not already happening. Will our current lifestyles influence our fertility as a species over time? What are the chances?

Then there is global warming to deal with in regard to the health and well-being of the elderly and the young when it comes to growth in the human population. Mothers have a tough time hydrating their babies and toddlers when the weather is too hot. Elderly people are also impacted by excessive heat, and we have had an example of that phenomenon in Australia when heatwaves hit with a vengeance and an alternative morgue, such as an ice rink, had to be quickly brought into action. Many in Australia depend on artificial cooling, such as air conditioners, because we are geographically a hot country, but with excessive amounts of energy needed to power them, the electricity grid can fail, as it has in the past. Global warming/heating plus changing climatic conditions will bring on more heatwaves and therefore heatstroke in vulnerable people; this is a given as the human species has also evolved to live in stable climatic/weather conditions. A classic analogy for the impact of global warming on all species is that the normal human body temperature is 36.7 to 37 degrees Celsius; if it gets to 38 degrees, then you have a fever-type temperature, but if you get to 40 degrees Celsius, then you are a hospital case because that is life-threatening. The big question is how hot is too hot for the human species to survive not just a heatwave but day after day, especially for billions of people who do not have the luxury of air conditioning?

Global warming will inevitably impact on the people of undeveloped countries who have not, in general, contributed to the environmental crisis, but they will also be exposed to extremes of temperature which they will not be able to escape from, so this is an environmental justice issue on a grand scale. The ethical question is do we care enough about our kin to change our ways so that they might still experience the climatic conditions their people evolved in and lived happily in for thousands of years?

We also know how susceptible the human species is to the transfer of animal and bird viruses, known as zoonotic diseases, especially given our experience of COVID-19. If we continue to abuse and exploit other life forms, then we will reap what we sow, as the good book says. Scientists are quite clear that as we take over more of the other-than-human natural world by way of deforestation and the depletion of grasslands which bring us into closeness with wild animals plus the consumption of wildlife, we are going to be exposed to more viruses that will jump from one host to another within the animal kingdom. We know the viruses and infections that have resulted from the other-than-human world are already extensive and well worth exploring to truly understand our relationship with the animal kingdom.

If we reflect on the statistics of the loss of life from COVID-19 around the world, it is possible to imagine that nature will take care of any human population explosion challenge the world could face in the foreseeable future, so exponential population growth may not be an issue. The worldwide death toll from COVID-19 must teach us something about the vulnerability of the human species to zoonotic viruses, and preparations by way of future vaccinations and quarantine facilities are already being considered in case zoonotic viruses become the new normal. If we do not make sense of our COVID-19 experience, then we are destined to live with zoonotic diseases.

Famine may be an existential threat to population growth as global warming and the changing climate continue to impact on the production of crops and livestock – that is, how much heat can plants and animals take before they succumb to global heating that is outside their evolutionary ability to survive? Can domestic animals survive three degrees of global warming? It is quite clear that there are many unknowns about population growth and how it will be influenced by food security because of the changing climate. What we do know is that Earth has the ability to sustain the population as it is now and the emerging population while we continue 'progress as usual' – but only with a monumental movement away from waste, widespread pollution, and the environmental degradation of life-supporting ecosystems. If we can get environmental sustainability under control, then we are in with a chance to provide sufficient food for an increase in human population and the population of all other creatures.

Then of course, there is the possibility of war to consider. It would appear that humans are creatures that are attracted to war with one another. We have designed the most extraordinary ways to kill one another, and with nuclear missiles aimed here and there, there is the potential risk of further destroying not only people but also the biodiversity of creatures we depend on, such as clean air and fresh water. We don't think that current wars in countries far removed from us will impact our health and well-being; it is the interlinked paradigm – that is, we are all connected as earthlings. Any activity that pollutes the air by way of warfare has consequences for all creatures.

There are also endemic health issues in our communities, with all kinds of cancers increasing exponentially. Breast cancer is a good example. One is left to wonder why breast cancer is on the increase since women's breasts are vital for feeding their newborns. There is also research done on the toxicity of breast milk, so we can only conclude that the problem of women's health lies in the air we breathe, the food we eat, the water we drink, the consumption of plastics, radiation from technology, and perhaps the cleaning chemicals we use in our households. There are most likely a myriad of reasons why human health, in so many ways, is impacted on by our way of life in the current environmental context. In the last fifty years, our lifestyles have changed dramatically; we thought our progress seemed like a good idea at the time, but now we are seeing that perhaps some changes were not in our best interest.

Some countries have aging populations, for example China and Australia. The young Australian population has been bolstered largely by immigration, which is creating a very multicultural Australia. The population explosion of babies after World War II are now coming to the end of their lives, and so the deaths of the elderly over the next ten years could help to flatten any exponential rise in population worldwide.

In summary, exponential population growth as an environmental concern is not a topic people like to contemplate, but it is another elephant in the room. Earth is just one planet, and it has to service the human population and also many billions of other creatures in perpetuity. Since the human species absolutely depends on other creatures, humans cannot go on taking the lion's share for themselves without depleting other creatures' life support systems that provide food and habitat security for them. The way forward is to be considerate about how we use Earth's gifts and to examine what it is that we need to do to maintain a healthy biosphere to ensure our longevity. Also, to accommodate an increase in population, we can downsize our demand on the gifts of Earth. We can do this in a thoughtful, ethical, and constructive way so that global population growth is not an environmental issue.

Environmental Refugees

Environmental refugees are faceless people for many. Not much thought is given to them as they lose their homes, land, and, for some, countries through anthropogenic-induced climate change exacerbated by global warming. Some environmental refugees are classified as climate migrants; however, they are people who have been forced to flee their homes and homeland because of gradual/incremental or imminent threats to their lives from water insecurity, drought, rising sea levels, or extreme weather events such as floods. Environmental refugees are real people, and as members of the Earth community, we must be prepared to welcome and embrace them as they are not the perpetrators of crimes against nature that we are guilty of; they are the victims of our polluting and consumerist lifestyles.

A good and clear example of potential environmental refugees are the people of the islands in the Pacific whose lands are under threat from rising sea levels. Kiribati is one of the countries that will be seriously impacted as its islands will be underwater by 2050 if not before. These are people who have just gone about their 'business as usual' at home for thousands of years until their homes began to take in water. Their homelands are so low lying that they will inevitably be absorbed by the ocean. They will be homeless, nationless, and stateless environmental refugees, and we as a global community will have to take on the responsibility to accept and rehabilitate them in whichever country they choose to make a new home.

It is not only the nations of islander people who will be at risk. Around the world, a number of cities are at risk as the ocean rises to new levels; even an increase of half a metre to one metre could mean evacuation for millions of people. A rise of one to two degrees Celsius in temperature in our homes is challenging but quite bearable, but when the global temperature rises by one to two degrees Celsius, it can mean an environmental emergency for many millions of people. People are at risk of becoming environmental refugees in their own countries. Alarm bells are not ringing furiously for the general public though because this projection is primarily based on an increase in rising temperatures. As the ocean heats up, water levels will be elevated as heat expands water. However, what could speed things up are the melting icecaps, and the big unknown is how fast they will melt, adding to sea level rise. Mother Nature does not have favourites, so every country must respond to the challenge to limit greenhouse gas emissions if we are to avoid millions of environmental refugees. Coastal areas of countries such as Bangladesh, India, Thailand, the Philippines, Japan, the Netherlands, Indonesia, and China are most at risk of some of their people becoming environmental refugees. Some coastal areas of Australia will not escape the challenge of ocean inundation either as the signs are already presenting themselves and making people nervous about the long-term safety of their properties.

The expectation is that there will be around three hundred million environmental refugees by 2050. My thinking is that this will be a reality for island countries long before 2050. This will be a humanitarian tragedy unfolding in real time; hence, people in developed countries who have big dollars at their disposal are buying up land away from potential coastal disaster areas in their countries as an insurance against their palatial coastal homes being absorbed by the ocean. It is insightful to observe what countries that understand global warming and climate change are doing. Their leaders are preparing long term for their people, while Australian leadership is soldiering on with greenhouse gas–emitting policies, thus playing environmental roulette for our people. Our government just does not respond to climate science at all and prefers to play word games like 'We have to keep the lights on' with a nominal agreement to meet emissions targets. In 2020, southern Queensland and outback New South Wales in Australia, there is already the potential of environmental refugees in our own country because of drought. Several towns have had no water, and it had to be trucked in daily for months.

We can't say that we don't know about the possibilities of environmental refugees from droughts or flooding. The potential for sea level rise is well documented; it is one of those 'domino effects' with 'feedback loops'. The double whammy is the melting of the icecaps, the Antarctic, the Arctic, Greenland, and glaciers. The melting ice increases the volume of ocean water with the potential to raise the ocean level over the twenty-first century, but this could be a challenge in the twentieth century as the melting of ice is already underway and will be exacerbated by the incremental heating of the land and ocean. This will mean many millions of displaced people to be housed, fed, and employed, not to mention the potential for conflict. My hope is that when this happens, my country's leadership will be big-hearted enough to be the humanitarian people Australians needs them to be. We as a people are justly proud of the fact that when the whips are cracking for help of any kind, Australians are there in the roll call. However, this is not a drill; the cards have already been played, and we have to wait and watch it unfold.

To halt the probability of the plight of environmental refugees, it is absolutely imperative to restore the balance and harmony of Earth's ecosystems; to do this, there needs to be a cultural change in our relationship with the natural world. We need to re-envision our future by raising our environmental consciousness, and this begins with grassroots education and action. We have to challenge the status quo of 'business as usual', which is consuming the gifts of Earth, and restore the right balance in our relationship with the rest of the natural world. Earth will continue to do what Earth does to be healthy, and we will have to humbly accept whatever comes for a while until we realise and accept that we are not in control. Mother Earth will continue to talk back until we collectively hear the cry of the Earth. In the case of our 'progress obsession', which is the mantra of governments, there must be a monumental change from dominance to respect for the total Earth community if we are to survive as a species. We have to be people on a mission, and right now, there is no greater mission than to walk lightly on the Earth and help to restore our relationship with the natural world that supports us if we are going to stop the possibility of millions of environmental refugees and avoid societal chaos as a result of refugees, food shortages and the breakdown of financial institutions.

Plight of Bees

Another environmental issue to be concerned about is bees. Insects and especially bees are vital to the pollination of plants, and they have been doing their job for millions of years. In a bee colony, every bee has a specific job to do in maintaining or protecting the hive, so they are very busy insects. They only live about a month; the queen is an exception. There are many different kinds of bees, but only the honeybee makes honey. The pollination of plants is their major contribution to a healthy Earth; whilst there are other pollinators, bees are thought to pollinate about 33 per cent of fruit, vegetables, and crops.

Bees are an interesting environmental sustainability issue. Worldwide, there is a growing bee shortage because of what is called *colony collapse disorder*, evidenced by the disappearance of bees. It seems that bees leave their hives and do not return, but the jury is still out on why they do not return. So what is happening to bees? Well, no one seems to know for sure, but scientists studying the decline in bee populations suggest several reasons. They are an increase in chemical use such as pesticides and herbicides that bees are exposed to while pollinating plants, destructive mite infestations, genetically modified crops, changes in pollen quality, the proliferation of mobile phones (which increases radiation through the air), global warming influencing the growth of viruses and parasites, climate change causing changes in weather patterns, and inadequate food supply through changes in land use. Any or all could be influencing bee population health. Some countries with diminishing bee populations have put bans on some agriculture chemicals.

The bee crisis is a recent phenomenon. Europe and America are deeply affected by bee shortages, and it is believed that half the bees are gone, but the bee colony collapse happened over an extended period, so their disappearance was not immediately noticeable. Scientists, while looking at all the possibilities, seem to think that a parasite in the gut of the bee is to blame. While honeybees seem to be most affected, there is concern that the parasite will spread to native bees as well. In Australia, the bee population seems safe for now, and bees are exported to countries where shortages have occurred. However, apiarists are on full alert because they know how valuable bees are and how vulnerable they are to any change in their environment.

From an economic point of view, the service bees provide in pollinating crops, vegetables, and plants cannot be replaced by human activity. Bees are and remain the principal pollinators, and therefore, billions of dollars in food production are at stake. Some people are pushing for a ban on insecticides and pesticides as a precautionary measure. What we can learn from the bee crisis is that the unexpected can happen when we fail to respect the needs of creatures in the natural world. It is worthwhile to look into a crystal ball to see what our food production might look like if we were to entirely lose the services of bees and other pollinating insects.

Natural Disasters

Natural disasters are somewhat misunderstood because natural disasters are part of the natural functions of Earth. As Planet Earth is alive, it simply does its thing as a planet, and so natural disasters are commonplace around the world. Natural disasters are Earth doing its business on a daily basis, and Earth makes no reference at all to the well-being of people or other creatures of the natural world. Earth doesn't do modelling of its behaviour or give media announcements; it just does what it does as part of its normal functioning as a planet.

'Biophobia' which means fear of the natural world, sometimes results from exposure to natural disasters as Earth can be seen as violent and dangerous in its day-to-day activities. As each natural disaster occurs, media televises it to the world, so people are bombarded with evidence of Earth's normal activities which often have severe consequences for nature, including us. Natural disasters are very different from human-induced disasters, such as a commercial plane crash or a nuclear energy plant explosion.

Natural disasters include floods, earthquakes, volcanic eruptions, landslides, droughts, tsunamis, cyclones, hailstorms, blizzards, heatwaves, cold snaps, avalanches, tornadoes, and bushfires. These are natural functions of Earth, and they have been going on forever. Natural disasters can cause the loss of life and huge property destruction, which is more consequential when they happen in built-up areas where millions of people have made their homes. Every year there are hundreds of natural disasters, and billions of dollars' worth of damage is incurred as natural disasters strike. It is therefore important to have a healthy respect for nature and make good choices with safety in mind.

As well as natural disasters and human-generated disasters, there is now another category of disasters to be examined, and that is how human activity is exacerbating natural disasters; for example, cyclones that may previously have been category three are now category four or five at landfall. At the same time, meteorologists are thinking that there needs to be a category six descriptor as cyclones are increasing their velocity. Scientists confirm that global warming causes the air and ocean to warm, which increases evaporation and, in turn, feeds the velocity of a cyclone. So what is historically a natural disaster, the influence of human activities can make the natural disaster so much worse. The prediction is that cyclones will get stronger and possibly more frequent as over the last ten years, Earth has experienced eight of the hottest years on record; 2019 was the hottest year on record, and already, some parts of the world are experiencing higher temperatures in their regions.

As I write this narrative, there are many wildfires in progress around the world. The use of the term 'unprecedented' is ramping up. It is used frequently to explain new phenomena within the realm of traditional natural disasters. As the Earth heats up, so does the atmosphere, hence more lightning-induced wildfires and more droughts that make forests tinder dry, so a mega wildfire is a consequence of a combined Mother Nature response. By that, I mean we have been told that there is a domino effect when Mother Nature decides to act. In late 2019 until early 2020, Australia had a rush of bushfires. Bushfires, we are used to, but these bushfires were described using that potent word – 'unprecedented'. Bushfires have a role to play in keeping forests healthy and fresh, but these bushfires not only burnt trees and forest vegetation but also scorched the Earth, and biodiversity had no escape. NASA photos reveal the 'before' and 'after' of the area destroyed by the bushfires. There is not much chance of the burnt hectares coming back as forests anytime soon, maybe never because the seeds in the ground that would normally rejuvenate the vegetation are also scorched, perhaps beyond germination, as they have done in the past.

Australia has always been a land of droughts and flooding rains, but global warming will inevitably make both more frequent and more severe. It is the same with bushfires. Australia is well experienced with bushfires, but unseasonal rises in temperature can turn a natural disaster into a holocaust for people, vegetation, and biodiversity. Bushfire seasons are already extended and are happening much earlier because of rising temperatures.

Rivers that were once able to cope with normal rainfall are now flooding towns and cities. Buildings, paved roads, and streets do not allow water to subside into the soil, and storm water drains are inadequate to cope with the severity of an increased volume during downpours of rain. The ocean, as another example, has always produced high tides and king tides, but low-lying land, especially islander people, will be more subject to flooding because as the ocean heats up, water expands, and sea levels rise, so the islands cannot cope with the new level of high tides.

To stop natural disasters from becoming so much more than they were originally as part of Earth's natural functioning, the cause must be addressed. We are one people of Earth, and we have to work together to change to alternative, renewable energy to stop polluting the atmosphere and live in the right relationship with Mother Nature. There is no other way to stop some natural disasters increasing in their velocity and their inevitable extreme destruction.

Alternative Energy

Manpower to horsepower to fossil fuel power to nuclear power to renewable power – the use of different power sources tells the story of humans' need for energy to drive progress into every new age. The exploration of alternative energy is a 'good news' story. Inventions to generate alternative power are still in the infancy stage but growing rapidly around the world. People are yet to realise a dream of a world powered by non-polluting sources, but I believe a new era is dawning because we are 'wise humans'.

If humans had not gone down the addictive path of fossil fuels, we would have a clearer idea of what the world could be like today. The mind boggles at the thought, but humans are creative inventors, and 'necessity is the mother of invention'. An excellent example of how quickly we could move to alternative energy is communication. When I was a child, our phone hung on the wall in a big brown wooden box. There was a receiver handpiece and a handle to turn to alert the local telegraph office that my mother wanted to make a call. The telephonist moved cords from one hole in a dashboard to another depending on the number to reach. An overseas call, highly expensive, had to be booked well in advance of the call time. Now we have mobile phones, which are unlimited libraries, entertainment arcades and offices in our pockets that instantly connect us around the globe. I find this analogy very comforting for some reason. All that alternative energy production needs to prosper is the will of the people to vote in a government that will legislate for innovation and change to alternative energy while backing it up with financial support to bring it to fruition. I think the doomsday oracles of 'the death of progress' or 'we will be left in the dark if we stop burning coal' have had their day. A new era is dawning, and we just have to hope that it rises in time to mitigate the worst impacts of greenhouse gases. Thankfully, we have a vote, and we can use it for the good of the natural world, which is good for us as well.

How fast can people move to alternative energy? The answer is very fast. All over the world, scientists, inventors, business entrepreneurs, and dreamers are designing new ways of being in a world that is responding to the needs of the natural world. Alternative energy sources are alternatives to fossil fuels to address diabolical pollution in the atmosphere and biosphere as well as destructive global warming. Alternative energy works with nature, capitalising on the free gifts of nature to generate energy from nature's gifts of wind, sunlight, ocean tides, rivers, and geothermal heat. These sources can produce an infinite amount of energy. Progress in the provision of renewable energy is well underway and moving very quickly to an adolescent stage of development. It is a very exciting era for explorers of innovative technology. The good and most positive thing about renewable energy is that it is relatively easy to produce when compared with the process of extracting fossil fuels. The most common alternative energy sources already proven to be effective are as follows:

1. *Solar* energy, powered by the sun
2. *Wind* energy, powered by the wind
3. *Geothermal* energy, powered from heat in the Earth
4. *Hydroelectric* energy, powered by water
5. *Biofuels*, powered from plant material (but not a long-term option)
6. *Nuclear* energy, powered by burning uranium (but this is not a viable long-term alternative because the management of nuclear waste is unresolved; no country wants nuclear waste in their backyard)

The sun, wind, earth heat, and water as sources of energy production are believed to be the best alternatives to fossil fuels because they will have the least impact on the environment. The production of biofuels comes at a cost to rainforests and cropping for food, so this is not sustainable even in the short term. Nuclear energy, while extremely efficient in producing energy, has a deadly downside, and that is how to manage toxic waste and the possibility of catastrophic nuclear plant disasters such as those in Chernobyl (1986) and Fukushima (2011).

Renewable energy targets are constantly being agreed on, and great progress is being made to achieve these targets, even if too slowly. The plan is that by 2050, we need to get to zero emissions, so the world's energy needs will have to come from alternative energy sources. This target timeframe is, of course, nowhere near good enough to stem the tide of global warming. Climate scientists argue that this target is too little and will be too late. The impact of global heating generating climate change, a growing population, and air pollution will drive commitments to set targets much earlier. The future of alternative energy is *now*. There is no time to lose because we have the knowledge and technological genius to advance with alternative energy at a cracking rate. What is lacking is the 'will' on the part of fossil-fuelled governments determined to literally hang on to dead dinosaurs. The discovery of exposed mammoths in Siberia because of melting permafrost is a warning to leave the dinosaurs of the past, their vegetation, and their habitats in the ground.

Could we let our imaginations loose for a minute and try to imagine what our world might be like had the human species not gone down the pathway of burning fossil fuels for energy?

Given that the human species is incredibly creative and innovative, what might be our situation now when it comes to the health of our home planet? We seriously need to dream of a new kind of wonder-world where people are united in their commitment to rid the Earth of any kind of energy provision that pollutes, live in unity as Earth ambassadors, and walk the talk in unison – that is, when humans and the rest of the natural world dance to the same tune. From my forty years of following the environmental disaster saga, *zero* emissions by 2050 will be *too late*. I would love, for the sake of my grandchildren, to be proved wrong. However, it is simply a matter of physics, mathematics, and chemistry. Photography can be the judge.

Alternative Farming

After looking at alternative options for producing energy, it is a good idea to get a fix on alternative farming. We depend on farms to produce our food. The 'slash, dig, and burn' days of farming are over from a philosophical and environmental perspective. The loss of previously fertile land has awakened the farming community everywhere, and many are now at the forefront of sustainable farming practices. Farmers around the world are subject to the changing climate through global warming, and they are clearly feeling the effects of it in some countries. Farmers are conscientious watchers of the weather, and they have begun to notice the changes in the seasons. Thankfully, today many of them are on the front foot to address the environmental issues that will challenge them and their farming practices into the future.

The majority of farmers in the world are 'subsistence' farmers who grow enough food for their families and livestock. However, in times of poor rainfall and subsequent drought, all can be lost. Farming is a risky business in the best of times, but there is so much less certainty today because of global warming and the changing seasons. Today, in Western countries, farms are thousands of acres in one holding, and they produce huge amounts of food to feed a growing population, most of whom live in cities and are completely dependent on farmers for the provision of all their food. However, a large percentage of crop farming, something like 70 per cent, is used to feed animals, especially cattle for milk, beef and chickens, so this may have to change in the future.

Numerous experiments with farming have taken place in different countries over time. Some farming techniques worked against nature and others with nature, producing very different long-term outcomes. Farmers today are highly skilled, and their businesses are now highly mechanised and computerised for planting broad acres of mono crops such as wheat, corn, rice, and soybeans using genetically modified seeds, fertilisers, chemical pesticides, and, in some cases, intense irrigation. One of the downsides is that fertilisers and all kinds of chemical products are poured onto the crops, which we know is detrimental to soil and soil microorganisms that are required to keep soil ecosystems healthy. Irrigation and rainfall flush these pollutants not only into the soil but also into creeks, rivers, wetlands, and all manner of waterways, and some find their way to the sea.

To this point, farmers are extremely successful in feeding the global population. However, there are grave concerns regarding the longevity of their success. Huge demands are being made on soil and water every season, and global warming influencing the changing climate is a new phenomenon impacting on the viability of farms. Grassland farms border deserts, and huge amounts of native grasslands have been taken over for cropping, so this has a genuine risk factor for the longevity of monocultures.

Farms have an opportunity to draw down carbon from the atmosphere and store it in healthy soil. This can be done by planting trees and using zero tillage to preserve life in the soil. It is the micro life in the soil that can play a huge role in sequestering greenhouse gases. This process is a natural way of drawing down carbon but one that has been overlooked because of changes in the way agriculture is being conducted – that is, dominance over soil rather than working with it for the long-term health of the soil.

As the effects of the changing climate are not felt yet in some farming areas, no doubt farmers will do their best to ensure their farms remain viable by tweaking and varying their traditional forms of farming. Two alternative farming methods are organic farming and permaculture. In a way, they both respond to the needs of Earth. Permaculture is designed for farmers who want to produce more food, relying on understanding the ecosystems and biodiversity on their farms. Organic farming, on the other hand, requires native seeds, natural fertilisers, natural pesticides, low impact on the soil, and the sustainable use of water. Both farming methods have been highly criticised for their inability to produce enough food for an ever-growing population. However, the farming practice is changing and must continue to change if it is to be respectful of the needs of ecosystems and biodiversity that are influenced by farming.

As farmers and farming are vital to feed the masses, we must have faith in them to do the right thing and be in the right relationship with nature. However, growing beef and producing dairy may not survive the challenges of the changing climate, but there is also the need to feed an increasing global population. We are, in general, addicted to meat and beef in particular. As the climate challenge unfolds, beef may have to depart the market. Why is that? It is commonly understood that cattle are responsible for about 10 per cent of greenhouses gas emissions, which is unsustainable. By comparison, aeroplanes produce about 3 per cent of human-induced pollutants in the atmosphere. When cattle burp and become flatulent, they release methane into the atmosphere, and methane is many times more potent as a pollutant. Just as CO_2 can be seen belching from coal-fired power stations, so can methane belching from feedlots be clearly seen from space.

Agriculture and cattle farmers are cutting and burning forests around the world, especially the Amazon Rainforest, to increase acreage for growing crops and grazing cattle. Those who know what the outcome of this behaviour will be are shaking their heads in absolute disbelief given their scientific knowledge of the role of the Amazon Rainforest in maintaining a healthy, balanced biosphere. Water for livestock is another issue, so cattle farmers are in a no-win situation and may inevitably have to give up their way of life, which, no doubt, will bring about much kicking and screaming from not only the cattle farmers but also us who love our meat. What the average consumer doesn't understand is that cattle farming has a huge water footprint as each cow or bull requires between twenty to seventy litres of water a day. Wow – that math is hard to chew on!

Nevertheless, Mother Nature could take care of this issue without the need for rallying and protesting as she regulates the rainfall and initiates droughts; stockfeed may end up in short supply globally as it did here in Australia for our beef producers who were regrettably brought to a halt by the ongoing drought. There is some good news that came out of this latest drought, and that was that traditional farmers who had braved droughts before were beginning to challenge the notion that droughts were all cyclical, to the realisation that there may be a connection between extended droughts and the changing climate. Farmers know the seasons are changing, and they are adjusting to that challenge.

The other good news is that some farmers are returning to traditional farming practices, such as rotation farming, to avoid natural grasses and soil on grasslands being denuded of nutrients or lost to erosion or salinity. Today the process is called 'regenerative farming', and the hallmark of this type of farming is to care for the soil first and foremost, followed by the importance of nurturing native grasses and replanting trees to hold the soil and bring back biodiversity so vital to healthy ecosystems. The health of soil and the health of biodiversity are absolutely linked to the health of crops or pasture.

As for agriculture that requires a huge amount of available fresh water to grow food for humanity, farmers are rethinking their practices in the light of seasonal changes. The growing of food may have to be done under shelters, such as huge greenhouses, as global warming continues to take its toll on agricultural production, such as grapes boiling on the vines. Without real change, the outlook for traditional farmers is not one to give a great sense of confidence for the future of food, especially if the weather and climate lose their reliability. Planting according to the traditional seasons is already a problem for many farmers. We are looking at less than a ten-year window of opportunity to reverse global warming impacting on the viability for traditional farming to continue around the world.

We are constantly hearing Australian farmers talking about coping with drought, which is, in one way, understandable because Australia is a very dry country, but something is different. It seems that traditional drought had an end, but there are some farmers who have yet to experience the end of drought. Given the human population of the Earth and every creature that needs to be fed, the outlook is somewhat unnerving for climate scientists and, indeed, agriculturalists. It should therefore be unnerving for us, who depend on them to produce our food.

CHAPTER TWELVE

Spiritual Response

Science versus Religion

In 1967, a historian, Prof Lyn White, caused an upset in the Christian churches when he published an article in the journal *Science* titled 'The Historical Roots of our Ecological Crisis' (10 March 1967, pp. 1203–1207), where he laid the blame for the environmental crisis at the feet of Christianity. White maintained that the biblical quote in Genesis 1 where God says to his people to 'subdue' the Earth and have 'dominion' over every creature, literally interpreted, has not only solidified a dualism between humankind and nature but also decreed that it is God's will that humankind utilise nature for their own purposes. He added "that Christianity bears an enormous burden of guilt for the environmental crisis and that it will continue to worsen until we discard the Christian axiom that nature has no other purpose than to be of service to humankind." White was severely criticised for his views in 1967 – that is, over fifty years ago – but we must agree with him to some extent given our continued subduing and domination of the natural world largely by Christian countries. Our 'use and abuse' mindset must have come from somewhere other than our ancestral hunters and gatherers who walked lightly on the Earth. However, it can be said that words like 'nature' and 'environment' suggest a separation of the human species from all other species, which has not helped our dualistic world view regarding our 'subduing and dominating' paradigm which continues today.

Science Invites All Religions to the Dance

Science and religion have never been really good bedfellows. There has been a distinct lack of trust in each other by both sides. However, the unpredictable occurred when thirty years ago, science and religion together formulated a request to religious leaders of the world to inform their adherents about a potential climate emergency. This would appear to be a strange combination of philosophical positions and very different from historical science versus religion antagonism. Scientists made their plea because major religions hold within their vision and mission statements a healthy respect for the natural world – that is, they accept nature as sacred, as spiritual, with a call to be faithful custodians of Earth.

All religions have within their faith beliefs, their values, their written texts, and their traditions an obligation to care for creation and all creatures. Scientists were passionately keen to get their research out to the masses, but they had no avenue for sharing their understanding of what was happening in the worldwide environment, so they chose religions as they are the organisations that could get their messages out to the general population. I think scientists have given up on religions of the world, but it must be said that there are an amazing number of faith communities that are proactive for the health of Earth today. It is not too late for ministers of religion to call their people to join the environmental movement to bring about change in the way we live on Earth. There is no more time, but just as the severity of COVID-19 took the world by surprise, the fear is that the climate catastrophe will, despite all the protestations of climate scientists over decades, still take the world by surprise.

The role of ministers of religion is to educate their people about what we are doing to our home planet, how to love our home planet, how to be resilient when the environmental emergency is acutely upon us, and the importance of loving and supporting one another as we cope with the crisis until we find a perhaps painful but necessary pathway to a new way of living in the right relationship with the rest of the natural world. The frustration of climate and environmental scientists is as palpable then as it is now. It is, of course, to our eternal shame that we have allowed their predictions for a climate emergency to come to pass on our watch. Even now, it can be said that we are, in general, still scurrying around in the dark, totally incompetent to realise the reality of the climate emergency before us.

It appears the human species is incapable of piecing together what is before us. No wonder climate scientists are teetering on despair, and for their own peace of mind, some have decided to give up on us and retreat into their laboratories. However, there are still some feisty fighters out there who are not watering the narrative down to make it more palatable for us. They are calling it out with a 'take it or leave it' attitude but knowing they speak their truth. Visible evidence of the climate catastrophe is so clear, we don't need climate scientists anymore; just a camera and a thermometer will do! What climate scientists predicted in 1990 is now coming to pass. Today religious leaders are just as exposed as everyone else to environmental information on the television, so they cannot be ignorant of what is occurring regarding global heating. We have camera action of what is happening to Earth, with unprecedented wildfires burning out of control, floods, melting ice, and deforestation and therefore the loss of biodiversity, people's lives, and, for other people, their livelihood, which gives everyone a reason to reflect on what is taking place.

Scientists invite you, more urgently beg you, to become proactive in your religious communities. It is important that we demand from our religious leaders that action on climate change become the social justice issue for faith communities as religious communities are called to hear the cry of the Earth and the cry of the poor.

Addressing global warming–induced climate change is the greatest spiritual, moral, and ethical challenge of our day. It doesn't matter what religious traditions we belong to as everyone is going to be impacted by the changing climate, perhaps not equally initially but, in the end, perhaps equally. Mother Earth has no favourites. For those of us who are in the Christian, Islam, or Judaic traditions, we will be answerable to God because it is God's Earth that we are destroying. Will God ask us where we were when the last bird or flying fox became extinct? If you read the Book of Job, Chapters 38 and 39, you may feel as I do that God will be questioning us with many 'Where were you when . . .' statements. No doubt other wisdom traditions have something akin to this responsibility.

Those in religious traditions have a twofold call to ecological conversion, first as earthlings and second as members of a tradition that believes God is the creator of all and that, therefore, all life belongs to God. For those in the Catholic tradition, Pope Francis has written a letter to 1.2 billion followers titled 'Laudato Si': Caring for our Common Home' (2015) to instruct and inspire action in people of goodwill. Pope Francis is all about integral ecology, which means it has to be part of everything we do. He says, 'All of us can cooperate as instruments of God for the care of creation, each according in his or her own culture, experience, involvements, and talents.' No doubt leaders of other religious traditions have also called on their followers to act for the health of our home planet, its people, and its biodiversity. No one is excluded from this call to protect and defend all life on Earth. In union with the plea of climate scientists to members of religious communities to take up the challenge to inform their people about the health of Earth, we can only wonder how different the world's response might have been to this climate emergency had religious traditions responded to the call from climate scientists in 1990.

Governments and people in general talk about 'environmental sustainability' as though everyone understands how to do this; also, environmental sustainability has pretty much been about environmental sustainability for the human species. Environmental sustainability efforts have an acute weakness in that it is not understood that you cannot jump from acknowledging that we are in an environmental crisis situation to environmental sustainability. There is, quite naturally, a process to go through; otherwise, efforts can flounder because often there is no head or heart to go with the hands. We need to know why we are in a crisis, how serious it is, and whether we can do what needs to be done in time to ensure our future; therefore, environmental sustainability requires an ecological conversion.

To come to an ecological conversion is a process:

- Global Environmental Crisis: We must acknowledge in our minds and in our hearts that we are in a worldwide environmental crisis and that we are responsible for it, that this crisis has now escalated to an environmental emergency pending environmental catastrophe.
- Environmental Awareness: This is the most basic position to hold. This is usually all about energy, recycling, and composting, but to move to the next levels requires education.

- Ecological Consciousness: This is a huge move as it requires us to understand our evolutionary story, our interrelatedness, our interconnectedness, and our absolute dependence on the health of the other-than-human natural world.
- Earth Spirituality: This requires a personal view in the context of our spiritual relationship with the natural world. It embraces our cosmic story; it is an experiential perspective of the Earth as spiritual, as sacred in itself. It acknowledges a sense of awe and wonder of the beauty, complexity, diversity, spirituality, and fragility of the Earth as a living unity of life. Earth spirituality is not restricted to religious traditions but transcends all boundaries of spirituality and science. It is a sense of oneness with the created world.
- Spiritual Ecological Consciousness: This movement encompasses ecological consciousness, Earth spirituality, and creation theology. It is a conscious acceptance that each person is spiritual and is most at home when experiencing connectedness to the Earth and God, who gives life and nurtures all creation. It requires an awakening of our capacity to connect spiritually with the macro cosmos and the micro of all living things as created by God and precious to God. It requires us to understand that creation is the first revelation of God.
- Ecological Conversion: This is achieved by immersing oneself in all of the above and sets us up to walk lightly on the Earth and be proactive for the health of all life on Earth. Once a person undergoes ecological conversion, there is no turning back to 'business as usual'.
- Ecological Restoration: This is all about rethinking our relationship with the biosphere and being prepared to restore, protect, nurture, and defend the rights of all creatures in perpetuity.

- Environmental Sustainability: This is the end result of the process – that is, once an ecological conversion has been undergone, the passion to sustain environmental sustainability will help power a new and awesome relationship with the natural world that will meet the demands and goals for environmental sustainability. Importantly, it means sustaining the whole planet – that is, its biomes, ecosystems, and biodiversity must be restored and sustained if the human species is to survive.

Earth Spirituality

We are the first generation of people to know the scientific history of the universe and that we, along with everything else, are participating in a very long and utterly marvellous cosmic story. Our fundamental relationship is with the natural world and, ultimately, with the universe. Each of us is the universe expressing itself in a concrete and unique way through our interrelatedness, our interconnectedness, and our absolute interdependence on nature. All social institutions and human societies reveal attempts to deal with this reality, and many books have been written to expound on our unique relationship with the natural world. We are children of the universe; we are made of the dust of dead stars. In a deep sense, we belong to the Earth; we are earthlings. This is a profound reality; there is no separateness from Mother Nature. If we understand our true place in the great story of the universe, then it will lend heads and hearts to the hands of environmental sustainability. Earth spirituality gives us a pathway to hear the cry of the Earth; it requires us to listen attentively to the voice of the Earth as she speaks to us.

Earth spirituality can be thought of as the oldest experience of spirituality. Indigenous people have much to teach us about our spiritual connectedness to the natural world. In their world view, they belong to the Earth and make no claim on owning the Earth or any of its biodiversity. Indigenous people know Earth as spiritual, Earth as sacred. We need them to be our mentors in the hope of reclaiming a sense of the Earth as sacred. It is also the most ancient spirituality in the Bible, so it is not new to us; we have just, quite simply, ignored our spiritual connection to the natural world as we carry on with 'business as usual'.

Our material success has brought us to a strange spiritual and moral bankruptcy with creation which has enabled us to plunder the Earth for our satisfaction. Our lifestyles have allowed us to negate our responsibility to care for creation that cares for us. Earth calls us to be quiet, to gently listen, and to reflect on our sacred Earth story as our physical health and mental well-being depend on spiritual connectedness to the natural world. Half the human population are physically separate from intimacy with nature. However, the human spirit seeks out the natural world, and we know from our own experience that when we are immersed in nature, we feel better, we feel stronger, and we feel in control of our lives.

If we reflect on our home planet from the perspective of astronauts who have viewed it from space, we can get a sense of the sacred, a spirituality of Earth, and be imbued with awe and wonder at the miracle, mystery, and majesty of our home planet. Astronauts have shared their experience and professed with great clarity that from space, there are no countries, no boundaries, just one beautiful blue dot that hangs like a fragile jewel against a backdrop of blackness. We are called to view our home planet with loving eyes. If we can learn to love our home planet, then we will always care for that which we love, so let's dedicate ourselves to loving our planetary home.

In summary, Earth spirituality, at its core, is all about sensitivity and reverence for the natural world. It is about feeling at one with nature. Earth spirituality is needed if we are going to commit to sustaining environmental sustainability because only love for our home planet will give us the strength we need to defend the natural world today and in perpetuity.

Spiritual Ecological Consciousness

My doctoral thesis was on spiritual ecological consciousness, which is an extension of Earth spirituality. It is especially attentive to the Judeo-Christian perspective of the sacredness of Earth. Sacredness is understood in this context as holy, as of special importance to God as the creator of all. In the Book of Numbers, Chapter 35, verses 33–34, it says that 'you shall not pollute the land in which you live . . . You shall not defile the land in which you live, in which I also dwell'. According to Judeo-Christian tradition, creation belongs to God, God dwells in his creation, and God is intimately invested in creation, and therefore, humans are subservient to the design of God for creation.

Humans are expected to honour, respect, and nurture the Earth because it is the first revelation of God. There is a covenantal relationship among God, his people, and his creation, as in Genesis 2:15: 'Yahweh God took the man and settled him in the Garden of Eden to cultivate and take care of it.' Our relationship with nature is imbued with the traditional call to stewardship of the Earth and the current call to ecological conversion because our relationship with our home planet has been separated, warped, and distorted, to the detriment of the total Earth community. Pope Francis has called us to awaken our relationship with the total Earth community, to take care of our common home through understanding and accepting integral ecology as a foundation for our relationship with the natural world.

Spiritual ecological consciousness is, therefore, the conscious acceptance that each person is spiritual and is one's most spiritual ecological self when experiencing connectedness to the natural world and the Spirit of God through understanding that we are a planetary species. Long-term sustainability requires a significant and impressive change in the way we live, but more importantly, what is needed is an essential moral and spiritual conversion. There is a strong belief by some scholars that the environmental crisis is fundamentally a spiritual crisis of huge proportions which has consequences for the Earth. For our own health and the health of the planet, it is important to reclaim a sense of the sacred, a spiritual approach to nature. There are many who believe that the environmental crisis is, first and foremost, a spiritual and moral crisis of disconnectedness to the natural world. This disconnectedness has allowed the human species to make huge demands on Earth's gifts to the point where many species of creatures are threatened or lost forever. In the field of psychology, it is argued that the repression of our ecological unconscious is the deepest root of our feeling of insecurity, our lack of cohesiveness in industrial societies. So nearly thirty years ago, there was solid thinking about human health and well-being that is synonymous with the health and well-being of the natural world.

The impact of our disconnection from the natural world is hardly mentioned in today's society, even though mental illness is a huge concern for many people. There is a great deal of research that has been done on the impact on human health in relation to our disconnect from the natural environment; even psychologists who treat mental illness can be suffering from the same malaise. Spiritual ecological consciousness that is based on faith in a creator (God) who cares for all creation brings all of the above into focus, and perhaps we will not resolve the environmental crisis unless we change our attitude towards embracing the sacredness of the natural world. It is not a case of getting in touch with our own feelings; it could be that as a society, we need to get in touch with our spiritual ecological selves. Perhaps then we will be prepared to modify our lifestyles and make good decisions for the health of Earth so the rest of creation can survive with us.

Ecological Conversion

Everyone is called to an ecological conversion if we are to stem the tide of global warming influencing the changing climate. Ecological conversion is a process, as Paul Collins in his book titled *God's Earth: Religion as If Matter Really Mattered* (1995). It points out that "we must pass through a real conversion process to come to an existential consciousness of our relationship to the world. It is only when we pass through this process that we realise that spirituality and ecology are not mutually exclusive but actually belong together". So the call to an ecological conversion comes from many sources: Mother Earth, climate scientists and activists, indigenous people, biodiversity, the youth of today, future generations, our own health and well-being, our own deep spirit, and religious traditions.

The pathway to ecological conversion is first to understand and accept that we truly are in an environmental crisis emergency. It is not good enough to be environmentally aware so that we make small changes to our lifestyles, such as recycling. We need to grow our ecological consciousness so that we are imbued with a sense of interrelatedness, interconnectedness, and absolute, total interdependence on the rest of the natural world. To do this, we can immerse ourselves in Earth spirituality to feel that oneness with creation that will give us stability and a deep desire to help restore the Earth. Give a tree a hug today.

For those who are believers in a living, loving God and who can draw on faith and mission to restore God's Earth, the invitation is just as valid today, so if all of the above is in place, then the next stage is to undergo an ecological conversion where everything we use and do must be with the health of Earth in mind. This process will begin the restoration of Earth, which will ensure environmental sustainability for all of Earth's community of beings and for future generations.

CHAPTER THIRTEEN

We Are All in This Together

Join the Dance

We have known for at least thirty years what the future would hold if we did not listen to the admonition of climate scientists and respond. What they forecasted is now coming to pass, so now that we know what is happening with the health of Earth, we are all invited to the dance. We have a most wonderful opportunity to get involved and bring about the change we want to see in our world. If you have read this book, then you obviously want to understand the complexity of the environmental crisis pending environmental emergency, so the next commitment is to join the dance.

It is commonly agreed that people will never spontaneously take action unless they receive the support of others who are joining in the dance. Sadly, governments will delay action until there is a critical mass of people who are prepared to dance to the tune of demanding action on the changing climate. We are at war with nature, so it requires everyone to collectively decide that a healthy Earth is worth fighting for; we must encourage educated people like yourself to mobilise across the whole of society, which is happening now with thousands of environmental action groups in every country, but speed is now of the essence. The dire warning of world scientists is now realised and absolutely confirmed by the latest state of the Earth report, and we are quite clearly in the eleventh hour.

Most people do not join the dance because they are unaware of the severity of the climatic effect on ecosystems and all biodiversity, so getting to the point of rebellion against environmental injustices has been a slow process. Everyone wants to be nice, but nice is not going to cut it now. Too late for nice! We have allowed the environmental emergency to creep up on us. Now we know what is happening and how we can fix it, so we simply must get about this most challenging work to bring about the changes that are now so desperately needed. This is, without a doubt, the greatest challenge in our time, but there are millions of people who are educated about the health of Earth who are prepared to put themselves and their freedom on the line to bring about action for cutting global emissions. While there have been thousands of different environmental groups started up out of concern for our environmental future, they have not had the necessary clout to bring about effective change. One such group that is making their cause felt is Extinction Rebellion (XR).

The modus operandi of XR's is 'non-violent/civil disobedience'. Countless XR activists from all walks of life, all ages, all religions, and all ideologies have taken up the XR call to non-violent protest, and many have been arrested and charged with civil disobedience. I am a member, so I know XR is not for the faint-hearted when you know you can be arrested and charged with civil disobedience, but grandparent rebels are out there in numbers and in the thick of rebellion because they want to support their grandchildren who are striking from school for 'Fridays for Future'. XR is not going to go away because they believe that the stakes are too high to not give their all for action on global warming. They will be doing annoying stuff, even dangerous stuff, and those still in denial will ridicule them and violently oppose their protests, but at the end of the battle, there will be consensus that 100 per cent action needs to be taken now because 'the never-never of 2050' will be too late.

Environmental action for climate change requires 3 to 4 per cent of people worldwide to join the dance before real constructive action will be taken by governments, businesses, and people in general. Feel free to join the dance of any environmental group, but to join the XR mob is to be a rebel and on the front line of action. To be an XR rebel is to believe that our environmentally safe future is a cause worth fighting for and a reason to get up each day and do something to deliver on climate action for all those you love. Joining the dance for climate action is not meant to be a hard slog, but rather, we should be delighted that we know what is happening to our home planet and that everyone is invited to join in the dance. I have not met a morbid environmental activist; they are optimistic, feisty rebels on a mission, and no one is excluded from their mission. Their code is that they are a community, so no one ends up in gaol without a lawyer standing by and someone to greet you when you exit from gaol with a food basket or celebratory drink. XR is all about rebelling against environmental injustice wherever it is found but equally about taking care of its members.

The majority of people don't think of themselves as having a role as climate change activists. The health of Earth is changing daily and calls us to address the changes regardless of how painful it can be for us. Some of us can't bear to contemplate the possibility of future environmental catastrophe or societal chaos of our own making, but it does change the way we think and behave, either positively or negatively, about what we hear about global warming and climate change. The reality is we can make a stand, or we can fold and let life as we know it become a challenge for survival. Mother Nature has a huge capacity to right our wrongs, but we have to make a stand with her now.

One thing to come out of COVID-19 is that biodiversity can and will respond very quickly if given half a chance. With all the shutdowns, lockdowns, and people restricted to iso, nature came out to play and put on a show of clear skies, cleaner rivers with fish, penguins in the streets, deer walking through towns, photographing villages that reveal the Himalayas in the distance – the list goes on. Mother Nature is ready to renew the Earth; we need to ready ourselves to join the Environmental Revolution, which is now in full swing, to bring all safely to an Ecological Period of the right relationship with the total community of beings. Finally, if you would like to help climate scientists in their work, you might consider joining a multi-faith environmental organisation, Faiths 4 Climate Justice at info@greenfaith.org, or any other environmental action group. They all need you!

For a potentially wonderful future, join the environmental dance so we can look back on this time, the greatest challenge for humanity, with pride in ourselves that we did all we could do as a collective 'power of one' to make a difference.

A Way Forward

With all that is going on with our home planet, it is believable why climate scientists are despondent and pessimistic as to whether we will be able to turn a potential catastrophe around if we continue 'business as usual'. Many of us know what is going on with the climate, and that is bad enough; it is what we don't know that is the problem, but climate scientists know. They know there are potential tipping points that could bring on large-scale, irreversible climatic events, and that is a great burden to them.

Throughout this book, I focussed on themes or refrains such as the following:

- our shared evolutionary journey;
- global warming/heating and the changing climate;
- the importance of taking care of all creatures to keep life on Earth in balance and harmony;
- that every creature that is lost to Earth's community of beings has had an important role to play, even if we don't know what it is that they did;
- that what we do to the other-than-human natural world, we do to ourselves;
- that biodiversity does more for the health of Earth than we do;
- that human life is the poorer every time a creature is lost to extinction from the Earth community;
- that time is of the essence as 2050 will be too late; and
- that climate scientists have been informing us of the potential of environmental collapse for decades.

When I started this book, I didn't think I would be writing a horror story, but I feel that is just what I have done as there doesn't seem to be very much going in our favour environmentally at the present time except for the 'power of one' people – that is, environmental activists, awesome entrepreneurs, business people, creative inventors and some state and local governments who are keeping global warming and climate change in focus for needing action. Sadly, we cannot depend on the media because they seem to go wherever the momentary wind takes them.

There is a story about the 'power of one' to make a difference. One could go to an iconic person like Gandhi, with his victory slogan 'non-violence, non-cooperation', but for people like you and me, this little story is more achievable. The story goes like this: a young boy was walking on the beach, and he came across a lot of little crabs that had been washed up on the shore. He picked up some of the live ones and put them back in the ocean. A jogger ran by and noticed what he was doing but kept jogging. Time passed, and the jogger was on his return run, but this time, he stopped and said to the boy, 'You are wasting your time. I have just run a kilometre or more, and the beach is covered in thousands of them. What you're doing won't make a difference!' The boy quietly picked up another crab that was still alive, put it back in the water, watched it swim away, and then said to the jogger, 'I made a difference for that one.' This is the best some of us can do; if each of us does something, then the collective whole can make a huge difference, even if it is retweeting an environmental tweet or picking up a plastic drinking straw or a cigarette butt so that it does not make its way into the ocean. That is a good thing to do. Have no doubt – this is the greatest work we can be engaged in because the human species has never had to face this crisis before, but now we as earthlings have a common purpose, and we all share in a common future. Climate scientists would say that everything we do to help, even small actions/decisions have value and will make a difference.

Many people have informed us about the Earth and its biodiversity. To acknowledge them all would require dozens of books. They are the people who excel in the 'power of one' to make a difference. I would like to mention one of those amazing people: Sir David Attenborough, an iconic environmentalist who has brought the wilderness and the wild into our lounge rooms over decades. He has given us a glimpse of the richness of the world through his panoramic camera of much of the biodiversity we may never see in person. Frequently, we are left to wonder about the genius of his photographers in capturing the most extraordinary images of the lived experience of every accessible creature, like antelopes and dragonflies. David Attenborough is like the ice; the natural world flows through his veins and fires his heart. He has a story to tell from a lifetime of capturing the wonders of creation, so if possible, try to view his films to get a clear picture of what we stand to lose as a species if we fail to protect the natural world. In more recent times, he has moved from the awesomeness and wonder of the natural world to very powerful notes of caution that all the majesty, magnificence, and mystifying beauty of the wild could be lost to us if we persist in our dominance over creatures in the wild. He is calling the entire world to action today as never before have humans so impacted on the health of the planet. Environmental action is the antidote to environmental anxiety; this is everyone's problem, and we are most definitely in this together. We can do this!

So what can I do?

1. Support climate scientists on Twitter and Facebook and use information technology to get their research out to people.
2. Stay in touch with the International Panel on Climate Change.
3. Take good care of nature that you are the custodian of in your backyard.
4. Educate yourself about our beautiful home planet. Stay abreast of current environmental science.
5. Live simply and walk lightly on Earth.
6. Immerse your children in the natural world.
7. Educate your children about how the Earth functions and teach them how to be resilient as this will be their best defence against the climate emergency.
8. Educate your children about the natural world and give them a sense of awe and wonder. Teach them 'ecophilia' – that is, to love their home planet.
9. Pepper your politicians whenever you hear or see something that is not working for the health of Earth. One letter is worth a hundred votes.

10. Grow your own garden of paradise with not only something healthy for yourself but also something for birds and insects.
11. Sort out your energy usage and see if you can downsize your footprint. Maybe you can save enough to install solar panels or build a greenhouse.
12. Join proactive environmental movement groups in your community. They all need your voice and support, and there are hundreds to pick from.
13. March and protest with the youth of the world.
14. If you really want to put yourself on the line, you might think of joining XR. They are a gutsy mob who really are disturbing the peace in an effort to get the attention of those in power who have the authority to make big changes for the health of the planet.
15. If you can bear it, bring up environmental topics with your friends. Be prepared for pushback though and don't take any negative response as personal. It is not!
16. When you see or hear something about the environment, investigate it. Begin your own diary and record what you hear on the media about environmental matters.
17. Sign up to a scientific website, such as www.sciencedaily.com, and tap on the environment so you don't get inundated with research from other areas of science. The scientists will keep you informed about what is going on with our home planet.
18. Search the internet to discover your local environmental climate scientists who are blogging about their work. Just tap on the environment though, or you will get swamped with everything worth reading!
19. Watch films that tell our human story leading to our climate emergency. There are plenty of films that are based on what is happening today or the possibility of future climate emergencies.

20. Especially watch Leonardo DiCaprio's film *Before the Flood*. He gives a comprehensive overview of what is happening to our world in real time. He actually went to the hotspots to see for himself so that he can affirm the science with his own eyes, and he doesn't have to quote anyone.
21. Every topic presented in this book can be researched by yourself to consolidate or update your knowledge. Research data is always being updated. At the time of writing this manuscript, for example, some species of whales were in serious decline, but now, because of climatic changes as well as less and later ice formation around Antarctica, they are increasing in numbers. Go figure that scenario!
22. If you are feeling the 'power of one', you might write to your local educational departments and ask them to make sure their students are being educated for the environmental world they will inherit.
23. If you are a home study parent or a teacher interested in the environment, you might like to have a look at English literacy/environmental studies worksheets for kids available on www.environmentaleducationworkbooks.com.
24. For those of a Christian persuasion, you might like to read Pope Francis's encyclical letter titled 'Laudato Si': Care for Our Common Home' (2015). It is very profound but very readable.
25. Maybe after reading this book, you might consider a reading circle of friends and discuss the contents together for solidarity towards a vision, mission, and action.
26. If you are really feeling the call to the 'power of one', then stand for public office and become an ambassador for the Earth.
27. For more up-to-date information about anything in the book, you might like to go to YouTube for an update on all the topics covered in this book for children and adults. Remember, education is a game changer for action to

stem the tide of climate catastrophe. For children, type in the words 'for kids' so it is presented in a safe way that children will understand and not too heavy.

We all come to our understanding of the environmental dilemma through different paths, but once we get the enormity of the crisis we are facing, we can stand together and be the 'power of one' to make a difference wherever we are on Earth. As an environmental educator, my thing is about education because 'education' changes everything; we are changed and empowered by education. As I pointed out earlier, there is a lot of talk about global warming and climate change, but it is so much more important to be talking about the loss of ecosystems and biodiversity which are affected by global warming.

We are not alone in our efforts to address environmental issues; there are environmentally active people in every country wherever they see a need to be proactive, and the momentum is growing as people become more environmentally conscious. If all of us could just educate one other person and that one other person could educate another, then before long, we would have that critical mass of people to bring about a renewed, wonderful way of being in the world. After you read this book, you might accidentally leave it on a bus!

CHAPTER FOURTEEN

Recapping the Endgame

Ambassadors for a Hospitable Earth

To draw this narrative to a conclusion, I need to go back to the beginning – that is, my concern for climate scientists. Climate scientists have been telling us for decades that we could face a climate emergency. Therefore, I am thinking of all the climate scientists who have been gagged if they speak their truth, not to mention their personal integrity is challenged, and hence, they can suffer from an ethical and moral conscience dilemma. It is because they have largely been silenced or their work discounted by ferocious and frequently vicious deniers that we are now in a climate emergency with a timeframe for our own self-destruction if we are unable to come together to turn it around. It is no wonder they are exasperated by inaction on the changing climate when the evidence is so clear.

Climate scientists are not supposed to have an emotional position on their research work. However, their work is unlike any other. Their work is to examine and professionally assess the health of Earth from an empirical scientific position. However, they are not emotionally immune to their findings as they are not studying something that will affect other people. It will affect them personally; they know all too well the predicted outcomes for our lives after so little action. In recent times, surveys have been done on the Australian public, and 80 per cent of people want action on the changing climate, but where will that action come from when we have a government that is in climate change denial and we have less than ten years to stop our current 'business as usual' to reverse the impact of global warming?

When you listen to climate scientists, watch their body language, and listen to their carefully crafted words, you get the feeling that they are holding back from telling the full truth. They are human; they feel like we do. Their knowledge is daunting for them, and who wants to be the bearer of bad news? We know how bearers of bad news in biblical times fared (remember the story of Jonah and the Whale; Jonah did not like the job description God gave him because of the consequences for those who went before him), and today it is no better. As they say, 'we shoot the messenger' rather than want to hear what they have to say. We find the truth about climate change and the potential consequences hard to get our heads and hearts around. Maybe that is why people prefer 'fake news'; it is more palatable, more digestible, which seems to be the order of the day. However, climate scientists, the prophets in our midst, have an urgency in their appeal to the public to put pressure on the powers that be to act for a hospitable Earth now; there is no more time! We are in the eleventh hour, and the climate clock is ticking awfully close to tripping midnight.

I cannot conclude this account of what climate scientists are so nervous about without mentioning Greta Thunberg, the very exciting phenomenon for climate activism. This young lady has been ridiculed, humiliated, and greatly maligned by many not-so-gracious media folks. She just popped up in Sweden one day, and now she is a leader for climate action when so many millions of young people so desperately wanted and indeed needed someone to lead their charge for a safer, healthier, and more hospitable world in the future. Time has passed since her School Strike for Climate protest while sitting on the pavement outside the Swedish Parliament in August 2018 at the tender age of 15. Greta has confessed that she has Asperger's syndrome, and she refers to it as her 'superpower'. Her age and gender as well as being on the spectrum should have worked against her, but no; those characteristics are witness to her strength to those who follow her lead in rebelling against climate inaction.

Greta reminds me of another young woman in history who fought for a cause, Joan of Arc. It could be imagined that Greta's weapon and armour is her Asperger's syndrome because it enables her to see the climate crisis and inaction to address the changing climate so clearly, and she is not caught up like the rest of us with emotional responses to the consequences of global warming. There is no way a climate scientist would say, 'You have to act as though your house is on fire because it is' or 'If there is action, I am hopeful. If there is no action, I cannot be.' People who have Asperger's are known to be very direct speakers on any issue. Neither is she disarmed by the tsunami of criticism she has received over her years of campaigning because in her Asperger's armoury is a lack of ability to respond with emotion.

Greta has presented a powerful case for climate action within the realms of power and is a great example of the 'power of one' to make a difference. All power to her and the youth she leads in demanding that action be taken to stop heating the planet, protect the ecosystems, and defend biodiversity supporting life! Fridays for Future is their clarion call to strike, and they have all over the world. If we older people cannot see the forest for the trees in relation to the changing climate and environmental degradation, then may the youth of the world hold their course and never waver until their goals are achieved! When COVID-19 is over, hopefully, we will see millions of young people back on the streets, proclaiming their potent protest messages as ambassadors for Earth. There is currently no greater cause to fight for. Those kids will not be stopped; they know their future is facing an inhospitable Earth, so they will make their demands to their governments, and I will be supporting them because I believe in student power more than I believe in adult power. We have let them down, so now we must stand with them.

In the early 2000s, I was teaching environmental studies within the religious education curriculum at my college. My teaching was initially a failure because no one was talking about the science of the potential loss of ecosystems and biodiversity through global warming and the consequences for the health of the Earth if we did not stop greenhouse gas emissions. After trial and error, I found a way forward, which was to teach about the Earth and how it functions from an 'awe and wonder' perspective. The tide turned for me, and it became the topic for my doctoral thesis, titled 'Spiritual Ecological Consciousness'. This process was very successful, and classes went from seven students to four classes over three years. In one of my classes, I had an 18-year-old student boy in Year 12 who said, 'Why have I been at school for twelve years and never heard about this before?' No one in education thought it was necessary back then to talk about the state of the environment and our part in its demise; environmental education was all about the sustainability of the human species.

While I am still all about the importance of teaching from an 'awe and wonder' paradigm as opposed to a 'gloom and doom' approach to environmental studies, time is of the essence to get the message out about the health of our home planet, and so the gloves are off as they say. We can no longer pussyfoot around with the reality of the climate emergency now. It is time for us to step into the light and take on the environmental action challenge. Many young people now get it; they seem to understand that we are in a climate crisis and that their future needs action now. My past students are now adults, so I am quietly hoping that what I taught them may be coming home to roost at last. I hope they are now ambassadors for Earth and ready to take a lead role for the environment in whatever career paths they are on.

In closing this chapter, I want to say something about my home country. Australia is known to be a land of drought, flooding rains, and bushfires, and we have had a very strong dose of all three in the last number of years. With each catastrophic event, the issue of the changing climate rises to the top of the conversation, and people cry out for action, but as the fires go out, rain returns to drought-stricken farmlands, and the floods subside, our societal conscience, environmental consciousness, and conversation move to addressing the consequences, the clean-up, the restoration of environmental damage, while the climate change call to action fades into the distance. Each environmental catastrophe seems to be followed up with some domestic, government, or societal issue, and the need to address global warming is once again put on the back burner for action. Climate activists do their very best to keep the global warming action flame alight, but it is just plain hard work.

Year after year, Australia is ranked worst or close to worst against other countries' commitments and actions in addressing climate change. Even with the call on governments to spend big dollars to help with the COVID-19 recession, green money is extremely scarce, which is another blow to our climate scientists; no wonder our climate scientists are distraught and constantly feel thwarted. It is a rude awakening that some countries are considering placing tariffs on Australian products because we are not doing our share to stop global warming, environmental degradation, and the loss of biodiversity. Our government is preoccupied with 'business as usual', the GDP bottom line, and budgets for votes. To recapitulate, the whole world needs a critical mass of people to actively demand climate action. There is so much at stake, so much to lose, and so little time to bring about the change necessary to turn our relationship with the natural world around.

CONCLUSION

The great work at this time that we are called to is to address our relationship with the natural world. Our consumerist and disposable mentality is completely out of whack with the needs of the natural world that supports us. In our arrogance, we are not cognisant of the finiteness of the gifts of the natural world, and Mother Nature has her limits of what can be sustained before environmental collapse. As I said in the beginning, this is not a matter of faith or belief; it is a matter of physics, chemistry, and mathematics that can be calculated and photographed. We must move swiftly away from the Anthropocene Era of dominance over the natural world to an Ecological Period of being in the right relationship with ecosystems and biodiversity that support our very existence. We must engage with the precautionary principle if we are to live in a wonderland as opposed to a waste land and make decisions that focus on the health of air, water, soil, and all the gifts of Earth.

As a species, we must examine our ecological footprint to avoid environmental bankruptcy by depleting or destroying ecosystems and losing precious biodiversity that have shared our evolutionary journey and are vital for life. A healthy Earth is conducive to healthy people as our well-being as a species depends on it. This generation must move out of the darkness of denial and inaction because we cannot continue to borrow Earth's resources/gifts from our grandchildren. With our knowledge, intelligence, and technology, we should be leaving a richer, more plentiful world to our grandchildren, not a depleted world and, in ten years without extensive action, an inhospitable world. Alone, we cannot change the course we are on, but together, exercising the collective 'power of one', we can make a difference. This is why we are all called to the great work of our time – that is, to dance to the tune of Mother Nature. With her help, we can do this.

If you dream it, you can do it! We know absolutely, if we do not address the changing climate to get to at least 45 per cent less emissions by 2030 and zero emissions by 2050 at the latest, what our world will be like. Climate science modelling and projections for environmental disaster are now much clearer than they were thirty-plus years ago because what climate scientists predicted back then is now coming to pass on schedule, so we can expect that their projections for the future will be just as accurate. The only difference is we are closing in on that timeline, and there is much greater urgency. The modelling that says we could get to three degrees of global heating is way outside our experience to survive. What if we could dream of a world of health and beauty for every creature and how we might bring it to reality? We are wise humans. As wise humans, we have the ability to do the most extraordinary things if we are absolutely committed to it. When there is a world war, we commit everything for victory; when there is a worldwide pandemic, we commit everything to its eradication. We are at war with nature. We are looking at an environmental catastrophe/apocalypse pending societal chaos. What if we were to put everything into the Environmental Revolution so that the other-than-human natural world could be victorious? Then we would all be winners. If we could just come together as a wise human species, we could change everything that is needed to restore the health of Earth and achieve equality for all.

We have constructed the world we live in where wealthy fossil fuel dealers are running our lives and ruining our home planet, so we can deconstruct it and replace it with a renewed, respectful relationship with creation. We must change the current system that is threatening our lives and future generations; there is no other alternative if we are to survive, and this could be achieved if enough people could believe that the human species is unlimited in our ability to dream and create a better world for all. All power to the dreamers who can envisage a new wonderland where all can celebrate and enjoy life as integral members of the total Earth community – as true earthling warriors!

I have brought up many environmental issues for you to think about, and above all, I have tried to join the dots for you. I have written a lot about global warming – very repetitive, you might say – but the plain fact of the matter is human-induced global heating affects everyone and everything everywhere. Nothing and nobody will be exempt from the consequences of global heating as everything on Earth is interlinked, and we are absolutely part of that linkage. We cannot physically live without the rest of the natural world, so we must protect it and nurture it back to health. We know our past, and we understand our present; now we have to envisage the future we want for ourselves, our families, our communities, and our world.

I have said repeatedly that we must be a voice for the Earth – a voice for its biomes, ecosystems, and biodiversity – but my loudest clarion call is that we be a voice for climate scientists and all their colleagues who are hindered from speaking for themselves. On their behalf, I beg you to listen to them and do the right thing in assisting them in any way you can in your backyards, neighbourhoods, and countries; the continuation of a healthy biosphere depends on it. As I have said before, my biggest fear is that even with forty-plus years of environmental science educating us and preparing us for the full consequences of our actions that have brought on global heating, it is still going to take us, as an earth community, by surprise. Will we be ready when the first tipping point occurs, causing a cascade of domino effects? I can only hope so.

Finally, COVID-19 isolation has given me plenty of time to reflect on my life's work and the future of life on Earth – that is, big metaphysical questions. As I sit on my deck, drinking coffee and feeling the gentle sun on my face while I look over my manuscript, I cannot see climate change, and I cannot feel climate change, so it is not happening for me. It is not within my experience; it is not my lived experience because I am sheltered from the reality of it. However, I follow the science, so I know it is true and happening around the world because I absolutely believe the science, climate scientists, and photography.

It is hard to get my wheels down as it has been over thirty years of listening, reading, watching, studying, and teaching adults and students about the possibility of global warming influencing climate change, environmental degradation, the spirituality of the Earth, and its attendant topics. If just one person like myself should read this book, then I thank you; it will have made my call to arms worthwhile. I would gladly have written it just for you. Please join me in the dance!

www.ingramcontent.com/pod-product-compliance
Lightning Source LLC
Chambersburg PA
CBHW020633220526
45464CB00001B/129